Born to Parse

Born to Parse
How Children Select Their Languages

David W. Lightfoot

The MIT Press
Cambridge, Massachusetts
London, England

The open access edition of this book was made possible by generous funding from Arcadia—a charitable fund of Lisbet Rausing and Peter Baldwin.

A charitable fund of Lisbet Rausing and Peter Baldwin

This book was set in Times New Roman by Westchester Publishing Services. Printed and bound in the United States of America.

Library of Congress Cataloging-in-Publication Data

Names: Lightfoot, David, 1945– author.
Title: Born to parse : how children select their languages / David W. Lightfoot.
Description: Cambridge : The MIT Press, 2020. | Includes bibliographical references and index.
Identifiers: LCCN 2019044359 | ISBN 9780262044097 (hardcover)
Subjects: LCSH: Language awareness in children. | Grammar, Comparative and general—Parsing. | Grammar, Comparative and general—Syntax. | Second language acquisition—Study and teaching (Elementary)
Classification: LCC P118.3 .L54 2020 | DDC 401/.93—dc23
LC record available at https://lccn.loc.gov/2019044359:

10 9 8 7 6 5 4 3 2 1

Contents

Preface

This book is about the acquisition of language by young children. For all children, acquisition begins very early, arguably in utero, when they often respond differently to the sounds their mother makes in her language than to sounds from other sources. Initially, for example, babbling happens with a wide range of sounds, drawn from all the languages of the world and possibly even beyond, and steadily narrows to the smaller number of sounds that the child will use for the rest of her life, selecting them for what we will be calling her internal language. There is nothing voluntary about this, no more than children decide voluntarily that it would be good to see in three dimensions. They just get on with it, growing their language system as their biology demands, developing the sounds they will use and the structures. Deaf children develop a gestural system and express their thoughts by that means. Much interesting work has been done on these early stages of acquisition, but this book will focus on what happens a bit later, in the extraordinary third year of life. In that year, syntax emerges rapidly and children develop into more or less full-fledged human beings. This holds regardless of whether the language users surrounding the child are speakers or signers, who express thoughts with equal richness.

Every parent has witnessed this apparent miracle. In that third year, children come to express and understand a wide range of thoughts in a language that they discover, select, and make their own, perhaps an individual, private form of what we call, say, Japanese or Javanese. Children experience and produce a finite number of expressions but they have the capacity to understand and use an infinite range. Once they can say something simple like *Heidi made that*, they might also understand similar structures like *Kirsten saw the movie there* and *Eric ate the cake when Alex drank OJ*. And so on, literally ad infinitum, in principle. Sometimes it is hard to make out the words but it gets easier as children fine-tune the sounds of their language and fine-tune the syntactic system in ways I will elaborate.

There is much for scientists to discover, many things that vary from one language user to another, but a central theme of this book is that, just as some birds are born to chirp, ants to follow a chemical trail back home, so humans are born to parse, born to assign fine-grained linguistic structures to what they experience. Everybody does it, normally at about the same age. These linguistic structures are drawn from a range of possibilities defined by invariant principles that have been discovered over recent decades and that are required to interpret particularities of the surrounding external language. Parsing is key.

There have been significant conceptual and technical shifts over past decades in our understanding of the abstract, INVARIANT principles of language, which are a function of what linguists call UNIVERSAL GRAMMAR (UG) and do not need to be learned (I use small caps for the first instance of major technical terms). Those shifts lead now to new approaches to parsing and to new analyses of VARIABLE properties. Variable properties show up in some but not all languages and do need to be learned; they differ from language to language, unlike invariant properties. We can take a new approach to language acquisition and there are good reasons to do so, because there are difficulties with our current notions of language variation and acquisition.

Children DISCOVER the structural CONTRASTS manifested by the variable properties in the ambient language, beginning with the contrasts among sounds or gestures on which their language system will be built. A little later, as they experience their ambient language, they begin to SELECT the first elements of their language system, which they need in order to understand what is said. Children parse with their emerging grammar. In parsing, they identify nouns, verbs, and other categories that may vary somewhat from one language user to another. They assign linguistic structures to the external language they hear, which TRIGGERS specific internal elements required for specific aspects of the parse. Children use what UG makes available, notably the binary-branching structures that emerge from recursive, bottom-up procedures called Project and Merge, plus what their emerging system, already partly formed, affords them. We used to call that system a grammar, but more recently we call it an I-LANGUAGE, *I* for internal and individual, emphasizing the fact that the internal system holds for an individual's brain and not for groups of people, like the group of all English speakers.

External language (E-LANGUAGE), on the other hand, is a very different kind of entity, language out there. It is what a child hears, and it is not structured, not discrete, nor represented in people's brains. E-language has no inherent structure but has structure assigned to it through parsing, after an initial I-language has begun to be triggered. If E-language shifts, for whatever reason, children may assign different structures, parse differently from earlier generations of language users, and thus attain a new grammar, a new I-language. What those new parses are and how they are selected by the innovative children provides information about acquisition and explains new variable properties emerging through new parses. Both E-language and I-languages play crucial, interacting roles: unstructured, amorphous E-language is parsed and an I-language system results.

There is no EVALUATION of I-languages and no PARAMETERS defined at UG. That, I will argue, has been an ill-chosen wild-goose chase. Rather, it is time to pursue a new vision of what variable properties are and how some are selected by young children, with UG open in ways I will describe. There is no miracle here; all children go through a similar kind of development, and we can achieve substantial scientific understanding of it.

We can have a more productive research paradigm if we abandon the search for parameters defined at UG, discard any evaluation metric, and dispense with positing an independent element of cognition known as a PARSER. That will represent a major simplification, reducing the machinery required for our theories and minimizing the information being attributed to our biology, along the lines sketched by proponents of the Minimalist Program. In eliminating UG-defined parameters, evaluation of grammars, and a distinct parser, I aim to make a substantial contribution to Minimalist ambitions.

Rather than seeing parsing as a processing "approximation" or "add-on" to the grammar, we will connect parsing more tightly to the grammar. This emphasis on integrated parsing echoes work by Bob Berwick, Janet Fodor, Heidi Getz, Marit Westergaard, Virginia Valian, Colin Phillips, and others. Phillips (2003a,b) works under the motto of "the grammar is the parser" and like him, I will dispense with an independent parser. Instead, I will focus on how children might use their internal language system to assign linguistic structure to what they hear, that is, to parse their ambient, external language. That enables them to select structures in their internal, private language. As a result, I adopt a new vision, at least within generative perspectives on acquisition: parsing is central to that vision, but it is implemented by the emerging language system, not by an independent parser.

There are some good case studies that point to a productive new paradigm, but we will also examine some difficult cases that show where new work is needed and how that new work should be conducted. We will focus on these difficult cases, on what is needed to improve analyses as our research paradigm is recast. Thinking in terms of parameters has led to much interesting work on variable properties, and that will need to be reconstrued in this new paradigm. Thinking through difficult cases where things remain to be discovered will be helpful for making the transition from one paradigm to another. I will be keen to show that under the new paradigm, we can have good empirical coverage as we make our radical simplifications.

Part of the pleasure of writing a book like this is to show people in related disciplines how linguists have come to analyze language acquisition and variation, in the hope that this will inform analyses of other areas of human cognition: memory, spatial cognition, emotion, and beyond. This book is written for philosophers, psychologists, neuroscientists, and linguists who see themselves as addressing questions of cognitive science broadly. A good case can be made that vision and language are two areas of human cognition where successful theories have been developed. Linguists have sometimes followed the lead of vision scientists, and understanding how languages may vary, how they are parsed by young children, may now cast light on the acquisition of other cognitive capacities and lead to deeper understanding. I shall include enough technical information to give the ideas real substance but shall use as little jargon as possible to make ideas accessible across disciplinary boundaries.

In putting together books of this kind, authors try out ideas in papers and lectures. Lectures include those given in classes at our home universities and some on the road. Readers familiar with my earlier work will recognize that the ideas here continue the commitment to examining "the logical problem of language acquisition" that I first formulated in *The Language Lottery* in 1982 and to "cue-based acquisition" from *The Development of Language* in 1999. Since then, ideas have changed significantly, as reflected in Lightfoot 2017b, as a result of the emphasis on parsing in a particular,

grammar-based way. Those changes were first explored in lectures
I gave at the Beijing Language and Culture University (BLCU) in
2015, where I learned from the lively linguistic community led by
Fuzhen Susan Si. In fact, China has played a significant role, thanks
to two visits generously organized by Ping Li, which also allowed
me to try out ideas at the 2015 Brain Science meeting in Shenzhen
and in 2018 at Jiangsu Normal University in Xuzhou, at Shanghai
Jiao Tong University, and again at BLCU. I also presented these
ideas in lectures at Newcastle University in the UK, including one
as part of their Insights public-lecture series, at the Universities of
Connecticut and Pennsylvania, and at the nineteenth Diachronic
Generative Syntax meeting (DiGS 19) in Stellenbosch, South Africa.
I am immensely grateful to those audiences and to the individuals
who followed up afterwards.

There is nothing quite like being able to sustain a thorough investigation over the period of a fifteen-week course, and I am indebted
to students in my Georgetown Diachronic Syntax class in spring
2019, who worked with a preliminary draft of this book. I am grateful to the three outstanding referees who advised MIT Press under
the editorship of Marc Lowenthal; they all knew my work over a
long period and understood how and why it had changed over recent
decades, giving me much helpful advice about how to knit my story
together and make it fit more coherently. John Whitman and Wal-
traud Paul helped me with the analysis of Chinese, and Heidi Getz
proved to be an invigorating coauthor when we put our work
together and came to understand the similarities in our analyses and
the differences (Getz & Lightfoot to appear). Heidi and I share a
view that linguists' reliance on parameters leads them to underes-
timate the richness of learning in the acquisition of language.
Enriching the learning involved by basing our analysis of it on parsing offers a good alternative to analysis through parameters.

Psychologist Betty Tuller has become a partner in so many ways
that it was natural for her to sharpen the ideas of the book and
make it more readable and accessible to people outside of linguis-

tics. Linguist Terje Lohndal started working on this approach to language analysis as a teenager and first wrote to me then, beginning a long-term, fruitful correspondence; both his work and the correspondence have helped to keep me honest in this book. However, these days publishers use advanced, digital facilities for their editorial work, which can be a challenge for those of a certain age. My standard response is to look for somebody under the age of thirty five. The young person who helped me navigate those challenges was Kate Kelso, who was masterful. I am enormously grateful to the wide-ranging intellects of Elan Dresher, Bill Idsardi, and Barbara Lust; they read the whole manuscript and gave rich commentary and much helpful advice.

It was a particular pleasure to be invited to sketch the ideas of this book at BLCU in 2018 at the inauguration of China's first department of linguistics, a major development for Chinese linguists, who build on hundreds of years of thinking about language. There I argued that studying language can be a good vehicle for teaching undergraduates about scientific investigation quite generally, as we did in the early days of the Department of Linguistics at the University of Maryland, which I helped to found in the early 1980s. Given the way our field has developed, researchers can lead students to generate productive, innovative findings early in their careers, without expensive equipment and without having to develop full command of a discipline like history, biology, or nuclear physics. Instead, students may take advantage of the laboratory in their heads, their knowledge of forms of Chinese in Beijing and of English in Maryland. The dynamism of Chinese academic life and the recent history of work on Chinese syntax will bring exciting developments, and I would like to think that this book might assist those efforts by exploring new ways of thinking about language acquisition and variation.

Work on the book turned into a family affair: I was helped by my sister-in-law, Sue Lightfoot, who prepared the index, and my daughter, Heidi Lightfoot, Founding Director of Together Design, who worked with MIT Press to produce the cover design.

1 Three Visions

1.1 Invariant Principles and Their Successes

The biolinguistic enterprise, seeking to find the biological basis of linguistic structures, reflects the work of many people from many countries analyzing many very different grammars. They have discovered a huge range of interesting, abstract properties, by pursuing a particular, Minimalist vision of what a grammar should look like. This is the first of the three visions of this chapter: the quest for simple, invariant principles.

Acquisition is the process whereby a child selects a grammar conforming to those invariant principles. Invariant principles are universal, restrictive, and appear to be common to the species, serving to explain the similarity of the internal languages of speakers of many historically unrelated languages. A grammar is what we used to call the formal, generative system that characterizes a person's mature language faculty, which is represented in the individual's mind/brain. A grammar is now often referred to as an internal language, individual language, or I-language. Grammars, I-languages—terms used interchangeably throughout the book— are subject to the universal, restrictive principles referred to above.

Rich, invariant principles have emerged, often in response to arguments from THE POVERTY OF THE STIMULUS, a notion I will discuss below. Such principles, defined universally, bridge the gap between information conveyed by a child's typically very limited experience and the rich information codified in mature grammars.

A simple example starts with the fact that in English, *wh-* elements occur at the front of expressions but may be understood in a wide range of positions, the strike-throughs:

(1) a. Who did you see ~~who~~?
    b. Who did you speak to ~~who~~?
    c. Who did you expect ~~who~~ to win?
    d. Who did you say ~~who~~ left town?
    e. Who did you say Kim visited ~~who~~?

We analyze this as *wh-* phrases being copied from the position in which they are understood to a fronted position where they are pronounced. But there are various positions in which a copied *wh-* item may not be understood, for example, following a complementizer *that*, as in (2a): **Who do you think that has telephoned?* (* indicates an expression that does not occur in people's speech). Further examples are in (2b–d).

(2) a. *Who do you think [that ~~who~~ has telephoned]?
    b. *Whose did she see [~~whose~~ pictures]?
    c. *Who did she wonder [~~who~~ left]?
    d. *Who did she meet the woman [who knew ~~who~~]?

Generally, children are viewed as experiencing, and learning from, simple, robust expressions that they hear. The fact that **Who do you think that has telephoned?* is not said constitutes NEGATIVE DATA, information that something does *not* exist—something that is, in fact, precluded by some principle, which is not learned. It is not learned because it cannot be learned, since it would have to be learned based on negative data available to an analyst but not to a two-year-old child. The two-year-old has no evidence for the restriction. Therefore, what a two-year-old hears, the stimulus, is not rich enough to fully determine what she comes to know. This is what linguists call the poverty of the stimulus, an important part of the logical problem of language acquisition.

Invariant principles explain negative data like (2), and it is postulated that they are available to children through their biology, that they are attributes of their genetic material, hence not learned, hence the solution to this poverty-of-stimulus problem. The principles explain how simple experiences can trigger rich structures in the biological grammars that constitute some form of Japanese or of Javanese (see Guasti 2016 for good textbook discussion of advances in language acquisition, escorting the reader from basic concepts to areas of current research in a theoretically well-informed fashion). Understanding the invariant principles that have been successfully identified illuminates how we might explain negative data and gain new ways to approach variable properties, properties that occur in some I-languages but not in others, an area where linguists have been conspicuously less successful (as we shall see in the next section).

However, here's an important point showing the need for abstract structures in the parses that children arrive at: the word-for-word translation of *Who do you think that has telephoned?* does exist in a number of languages, as noted by Rizzi 1990: §2.6. This means that those non-English forms must have a different abstract structure than the nonexistent English forms, a structure that is the result of a child's parsing. Superficially similar sentences can derive from structures that differ in crucial and nonobvious ways; that fundamental point is not sufficiently appreciated even by some linguists. Thus, for example, Italian *Chi credi che abbia telefonato?*, literally 'Who do you think that has telephoned?', has an independently motivated (partial) structure *Chi credi che ~~chi~~ abbia telefonato ~~chi~~?*, where the embedded subject DP *chi*, 'who', is first copied to the post-VP position indicated by the second ~~chi~~ and then copied again into the matrix clause. English I-languages do not copy subject DPs to post-VP positions, but Italian children can learn to parse structures with a post-VP subject DP by hearing and understanding something simple like *Gianni crede che abbia telefonato Maria*, 'Gianni believes that Maria has telephoned', or even simpler *Ha telefonato Maria*, 'Maria has telephoned'. English-speaking children hear no such forms and therefore do not understand or produce expressions like *John believes that has telephoned Mary* or *Has called Mary*. The ambient language does not trigger such inverted expressions, which are therefore not generated by the emerging grammar, unlike what happens for Italian children.

Two universal, invariant properties of all languages that are not learned are recursion and compositionality. All I-languages seem to have three recursive devices, looping functions that allow the repetition of clause types; the existence of these recursive devices means that humans have, in principle, the capacity to generate structures of indefinite length. The three are relative clauses, illustrated in (3a), complement clauses, in (3b), and coordination, in (3c).

(3) a. This is the cow that kicked the dog that chased the cat that killed the rat that caught the mouse that nibbled the cheese that lay in the house that Jack built.

   b. Ray said that Kay said that Jay thought that Fay said that Gay told…

   c. Ray and Kay went to the movie and Jay and Fay to the store, while Gay and May worked where Shay and Clay watched.

Second, I-languages are compositional; structures are binary branching and consist of units that, in turn, consist of smaller units, which consist of still smaller units. So *saw a man with binoculars* may have the structure of (4a). In that case, since *a man with binoculars* is a DP constituent, the meaning is 'saw a man who had binoculars'. Another possible meaning of the phrase *saw a man with binoculars* is 'saw a man by using binoculars', in which case the expression has a different structure, the one in (4b). Here *a man with binoculars* is not a single constituent as in (4a); instead, the preposition phrase *with binoculars* is generated as an adjunct to the VP *saw a man*.

(4a)   (4b)

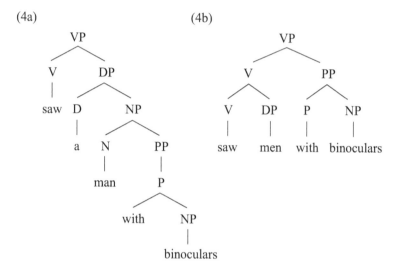

The 1970s saw the formulation of very specific conditions on grammatical operations that disbarred long-distance operations taking place across certain intervening elements. The Subjacency Condition restricted movement to local relationships (accounting for the nonoccurrence of forms like *What did she wonder who bought?*). The Tensed-S Condition ruled out nonoccurring forms like *They expected that each other would win,* and the Specified-Subject Condition eliminated *They expected the women to visit each other* on the reading where *each other* COREFERS with *they* (it may of course corefer with *the women*). These conditions disbarred operations across certain intervening elements. This was progress, but there were dramatic simplifying effects when scientists reformulated conditions to make analyses more readily learnable. For example, Lasnik 1976 rejected earlier attempts to formulate operations specifying what pronouns could corefer with in preference for specifying what pronouns could *not* corefer with. Lasnik turned things around in a way that led to a far simpler and more readily learnable account of the referential properties of pronouns.

The 1980s opened with Chomsky 1981a, focusing on the central and general role of GOVERNMENT relations and the spectacular simplifications stemming from the BINDING principles. This replaced the complexities of Chomsky 1980, particularly the stipulations of the indexing conventions given in the appendix to that paper. Chomsky built on Lasnik's reformulation and provided elegant accounts of binding relations, which we will examine in detail in §4.1.

Work during this period greatly enriched ideas about what kind of information might be built into our biology, as invariant principles of UG. Those principles enable children to go through that extraordinary third year of life, when, on the basis of rudimentary experience, they develop into recognizable human beings, thinking, understanding, and speaking more or less like an adult over an infinite range of thoughts, or at least like a person who will eventually develop into an adult human being.

My goal here is not to have readers relive the 1970s and learn how various principles were discovered. But readers do need to have a sense of the broad nature of that early work and of what kinds of problems it solved. I am providing some detail but not enough to be comprehensive. That would be a different book, which, having lived through the 1970s, I am not ready to write now.

One example of an early-formulated invariant principle is a condition on DELETION operations: that they can affect only an element that is in a prominent, easily detectable position, namely as (or inside) the COMPLEMENT of an overt head that is adjacent to that complement (see Lightfoot 2006b). So the complementizer *that* may be deleted in (5) to yield (6) but not in (7) to yield (8). In (7) *that* is not the complement of nor in the complement of the adjacent word; in (7a), for instance, [*that Kay wrote*] completes the meaning of the nonadjacent *book* but not of the adjacent *yesterday*.

(5) a. Jill said [that Jane left].
  b. The book [that Jill wrote] arrived.
  c. It was obvious [that Jill left].

(6) a. Jill said [Jane left].
  b. The book [Jill wrote] arrived.
  c. It was obvious [Jill left].

(7) a. The book arrived yesterday [that Kay wrote].
  b. [That Kay left] was obvious to all of us.

(8) a. *The book arrived yesterday [Kay wrote].
  b. *[Kay left] was obvious to all of us.

This condition on deletion also distinguishes the (a) and (b) examples in (9) and (10): the deleted (empty) VP in (9b) and (10b) fails to meet our condition, since it is not adjacent to the word of which it is the complement, *had*, whose meaning it completes; it is adjacent only in the (a) structures.[1]

(9) a. They denied reading it, although they all had $_{VP}e$.
  b. *They denied reading it, although they had all $_{VP}e$.

(10) a. They denied reading it, although they often had $_{VP}e$.
  b. *They denied reading it, although they had often $_{VP}e$.

Over the last two decades or so, under the Minimalist Program, there has been a change of emphasis: linguists have sought to simplify the principles, "minimizing" the information they embody. One form this has taken is to adopt an architecture that subsumes certain apparently distinct principles. For example, in the 1960s and 1970s linguists wrote very specific top-down phrase-structure rules to capture initial "deep" structures, which included complex structural properties that were, *ex hypothesi*, learned by children. Chomsky's *Syntactic Structures* (1957: 39) sketches a phrase-structure rule Aux → C (M) (*have* + *en*) (*be* + *ing*) (*be* + *en*), which contains much language-specific information, even specific English morphemes. Now things are very different: there is a general procedure for building structures, whereby a single, simple, recursive operation of MERGE creates binary-branching hierarchical structures. The invariant computational operation of (internal and external) Merge builds hierarchical structures bottom up. These structures combine heads with complements and phrasal categories with specifiers and adjuncts; this applies for all languages. This repeatable operation assembles two syntactic elements X and Y into a unit, which may, in turn, be merged with another element to form another phrase and so on. Merge is defined as Merge(X,Y) = {X,Y} and thereby derives Third-Factor No Tampering, Inclusiveness, and the restriction to binary-branching structures (for discussion of third-factor elements, see Chomsky 2005). This means that as two elements, X and Y, are merged into a third category, neither of the two merged elements may be changed in any further way as a function of that operation. As Epstein, Obata, and Seeley 2017: 482 puts it, "by definition, neither X nor Y is altered by the operation and no new features are added to X or to Y in the constructed object, nor are any deleted from X or from Y." Hence apparent properties of UG are derived through the invocation of the simple Merge operation.

Elements are drawn from the lexicon and merged into structures one by one; Merge is the fundamental structure-building operation. To clarify, the verb *visit* may be merged with the noun *London* to yield a VP, $_{VP}[_V visit \ _N London]$, but that also shows the effects of PROJECT, a distinct aspect of structure building, because the verb *visit* projects to a verb phrase, VP. Then the Inflection element *will* can be merged with that VP to yield an IP, projecting from Infl: $_{IP}[_{Infl}will \, _{VP}[_{V}visit \, _{N}London]]$. Then the (pro)noun *you* can be merged with that IP to yield another IP: $_{IP}[_{N}you \, _{IP}[_{Infl}will \, _{VP}[_{V}visit \, _{N}London]]]$.

An expression *What did you buy?* is built bottom up in the same way. At a certain point, the IP *you did buy what* has been built: *buy* merges with *what* to yield a VP, then *did* is merged with the VP to yield an IP, and then *you* is merged with the IP to yield another IP, as above. Then what happens is that the previously merged element *did* is copied and merged again, and *what* undergoes the same process. In both cases, the copied element is later deleted in the original position from which it was copied, as indicated by the strike-through: *What did [you ~~did~~ buy ~~what~~]*. Under this approach, there is no primitive operation of movement as such; rather, a copied element may be merged and then subsequently deleted.

Repeated application of Merge is the engine of the computational operations that relate, for any expression, the phonological and semantic forms, which are interpreted at what are commonly called these days the sensorimotor ("articulatory-perceptual" in Chomsky 1995) interface and the conceptual-intentional interface. These interface forms still must meet their own requirements, as we will explore in chapter 4 (for discussion, see Chomsky 1995: 168–170).

Minimalism envisions general principles that are not learned but that limit structures and operations in such a way as to permit learnable operations that express, for example, the former specificities of phrase-structure rules. It is worth emphasizing here that our focus in this section lies in the way that Minimalists have sought to simplify and minimize the information in the invariant principles being attributed to UG and to human biology. Our goal has not been to catalog everything in UG but we do need to recognize that the invariant principles identified have consequences for what we can postulate in individual I-languages, after learning has taken place. For example, consider Chomsky's Aux rule from *Syntactic structures*, referred to four paragraphs ago, Aux → C (M) (*have + en*) (*be + ing*) (*be + en*). Heads such as tense markers (C), modals (*will, may, can, shall,* etc.), and aspectuals like *have* and *be* merge with their complements sequentially, and the particularities need to be learned by children as they parse their ambient, external language and discover its elements; we will discuss how this happens in §2.4, when we consider how these elements first emerged in the history of English.

Similarly, the conceptual and sensorimotor interfaces have their own properties, which may also capture language-specific, learned, variable properties. Den Dikken 2012 is a comprehensive compendium of developments in generative syntax over the last several decades, beginning with rich, complex transformations that developed into very general operations like Move α, or even Affect α. One sees in children the steady emergence of the simple Minimalist operations, with specificities emerging from interaction across operations. Nowhere is this clearer than in the limits on which DPs may corefer and which must be disjoint in reference. The very complex indexing conventions of the 1960s and 1970s have now given way to the simple and general binding principles to be discussed in chapter 4, under which children must learn which words are anaphors and which are pronouns, apparently a feasible task.

Part of the motivation for minimizing the information in UG is the legacy of William of Occam's simplicity in theorizing, always seeking simpler and therefore more beautiful analyses. Another part is the goal of providing a plausible biological account whereby we might attribute the evolution of the language faculty in the species to a single mutation at some level. Taking Merge to be a universal, invariant property raises the prospect that the possibility of Merge was the mutation that made language and thought possible for *homo sapiens*. Berwick and Chomsky (2016) showed why such a view might be productive and elicited a judicious and informed review from paleoanthropologist Ian Tattersall (2016).

Thinking in terms of simple hierarchical structures resulting from Minimalist computational operations, notably Merge and Project, has also informed remarkable neuroscientific work linking brain activity to the abstract structural units underlying language and thought. For example: as we discussed earlier, repeated application of Merge guarantees that syntactic structures are binary branching and therefore involve a narrow range of hierarchical relations. That has suggested for many years that children instinctively parse expressions in terms of those hierarchical relations and not in terms of purely linear sequences. Now we have neurophysiological evidence that this is so, suggesting that universal aspects of the structure of languages correlate to some degree with a predetermined brain system.

In an experiment reported in Musso et al. 2003, Andrea Moro and colleagues exposed German speakers who had no previous encounters with Italian to an artificial variety of that language with Italian words but some variable syntactic properties different from those of natural Italian; they did the same with another group of German speakers and an artificial variety of Japanese. So, for example, there were no Italian-style null subjects: people would hear *Io mangio una pizza*, 'I eat a pizza', and never the subjectless *Mangio una pizza*, contrary to what occurs in normal, native Italian. The native German speakers were also exposed to "impossible" variable properties of Italian/Japanese, for instance, the negative marker placed after the third word in the unstructured expression, as never occurs in natural languages. So both groups of German speakers were exposed to naturally possible and naturally impossible artificial varieties of each language, Italian and Japanese.

The investigators analyzed their subjects' behavior and tested the brain activity of those acquiring possible and impossible artificial languages. Subjects learned the real and unreal-but-possible languages with similar accuracy. fMRI results, however, showed significantly different brain activity in Broca's area and elsewhere:

there was a "correlation between the increase in BOLD [blood-oxygen-level-dependent] signal in the left inferior frontal gyrus and the on-line performance for the real, but not for the unnatural, impossible language learning tasks. This stands as neurophysiological evidence that the acquisition of new linguistic competence in adults involves a brain system that is different from that involved in learning grammar rules that violate UG" and is based entirely on nonhierarchical, linear order (pp. 777–778). The authors go on: "[a]ctivation of Broca's area is independent of the language (English, Chinese, German, Italian, or Japanese) of subjects,[2] suggesting a universal syntactic specialization of this area" among natural languages (conforming to principles of UG) (p. 778).[3] "Our results indicate that the left inferior frontal gyrus is centrally involved in the acquisition of new linguistic competence, but only when the new language is otherwise based on principles of UG. The anatomical and functional features of Broca's area allow us to speculate that the differentiation of this area may represent an evolutionary development of great significance, differentiating humans from other primates" (p. 779).

We remain far from understanding the neurophysiology of the language faculty and far from knowing the neural mechanisms for processing hierarchical syntactic structure, but these results do suggest that when the language faculty is "switched on," producing or understanding language, certain kinds of brain activity are involved that are not involved when dealing with different, nonlanguage events.

As further illumination, David Poeppel and colleagues showed that when people listen to connected speech, cortical activity of different timescales tracks the time course of abstract structures at different hierarchical levels, such as words, phrases, and sentences (Ding et al. 2016). There are some problematic aspects to this study, but results indicate that "a hierarchy of neural processing timescales underlies grammar-based internal construction of hierarchical linguistic structure" (p. 158). Ding et al. found neural activity that directly reflects the abstract structures that linguists have postulated for the infrastructure of language, needed to account for the way that expressions are understood and used. See also Nelson et al. 2017. We always knew that the brain would need a mechanism for encoding the abstract structures of different levels, and now we can begin to figure out some of the brain activity that takes place when that mechanism is operating, a major development.

Ding et al. discovered that the brain tracks units at each level of hierarchical structure simultaneously. Such tracking requires knowledge of how words and phrases are structurally related. Heidi Getz et al. (2018) also asked how neural tracking emerges as knowledge of phrase structure is acquired in an artificial language. They recorded electrophysiological data (magnetoencephalography) while adults listened to a miniature language with distributional cues to phrase structure or to a control language without the distributional cues. They found that neural tracking of phrases developed rapidly when participants formed mental representations of phrase structure, as measured behaviorally, thereby illuminating the mechanisms through which abstract mental representations are acquired and processed by the brain.

Minimalist ideas about hierarchical structures being formed by multiple applications of Merge not only help us think differently about the evolution of the language faculty and of thought in the species and stimulate new neuroscientific work; they have also facilitated new approaches to the acquisition of language by children. The hierarchical structures formed by multiple applications of Merge constitute the means by which people, including very young children, begin to analyze and parse what they hear—the key component of the acquisition process, as Janet Fodor argued long ago in an important pair of papers, "Learning to parse?" and "Parsing to learn" (Fodor 1998b,c). To parse is to assign linguistic structures to I-language units and their interrelationships. We may now be at the point where we can dispense with independent parsing procedures, a goal to which Colin Phillips has devoted much of his career; see Phillips 2003a,b. Given the way in which hierarchical structures are built, we might argue that assigning structures to expressions is simply a matter of using the binary-branching structures that UG (in particular the Project and Merge operations) makes available and the structures that children discover as they parse their ambient E-language. Under this view, parsing is not a function of nonlinguistic conditions. Rather, parsing is just a function of the emerging I-language (as we will explore in chapter 3).

Work has shown repeatedly that children rely on the tools provided by their biology and learn much from little experience. Research has examined language acquisition by children exposed only to unusually restricted data, much of the work focusing on the acquisition of signed systems. A striking fact is that 90 percent of deaf children are born to hearing parents, who are normally not experienced in using signed systems and often learn a primitive kind of pidgin to permit rudimentary communication. In such contexts, children surpass their models readily and dramatically and acquire effectively normal mature capacities, despite the limitations of their parents' signing (Newport 1998; Hudson Kam & Newport 2005; Singleton & Newport 2004).

This is not surprising in light of studies of creoles more generally (Aboh 2017) and of new languages beyond creoles, which show that children exposed to very limited experiences go well beyond their models in quickly developing the first instances of rich, new I-languages (Lightfoot 2005, 2006a). Not much is needed for a rich capacity to emerge, as documented by many contributors to Piattelli-Palmarini and Berwick 2013 and now by Belletti 2017. Belletti offers a new kind of poverty-of-stimulus argument, showing that children sometimes overextend certain constructions, using them much more freely and creatively than their adult models.

Extraordinary events have cast new light on these matters: the birth of new languages in Nicaragua and in Bedouin communities in Israel. In Nicaragua the Somoza dictatorship treated the deaf as subhuman and barred them from congregating. Consequently, deaf children were raised mostly at home, had no exposure to fluent signers or to a language community, were isolated from each other, and had access only to home signs and gestures. The Sandinistas took over the government in 1979 and provided a school where the deaf could mingle, soon to have four hundred deaf children enrolled. Initially the goal was to have them learn spoken Spanish through lip reading and finger spelling, but this was not successful. Instead, the schoolyard, streets, and school buses provided good vehicles for communication and the students combined gestures and home signs to create first a pidgin-like system, then a kind of productive creole, and eventually their own language, Nicaraguan Sign Language. The creation of a new language community took place over only a few decades. This may be the first time that linguists have witnessed the birth of a new language *ex nihilo* and they were able to analyze it and its development in detail. Kegl, Senghas, and Coppola 1998 provides a good general account and Senghas, Kita, and Özyürek 2004 examines one striking development, where certain signs proved to be not acquirable by children and were eliminated from the emerging language, an interesting example of where adult learning differs from that of children.

Sandler et al. 2005 discusses the birth of another sign language among Bedouin communities in Israel, which has arisen in ways similar to Nicaraguan Sign Language and was discovered at about the same time. These two discoveries have provided natural laboratories to study the capacity that children exposed to unusually limited linguistic experience have to go far beyond their models and attain mature I-languages that conform to general, known principles.

Now Rachel Mayberry has discovered a third new sign language (Mayberry & Kluender 2018). In March 2016, thirty-five deaf children were brought together in two all-deaf classrooms in Iquitos, Peru. They were identified as still using home signs, before their new, common language began to emerge. The goal of Mayberry's work was to understand the very first stages of language emergence. In contrast, the first work on early speakers of Nicaraguan Sign Language was done ten years after the common language had begun to emerge, so much development had already taken place. Mayberry was able to gather data about deaf children's gestures before they met other deaf children, studying how their gestures changed throughout their first year of interaction with each other. Deaf children across the world often grow up without access to a shared language, so Mayberry's data will provide the first documentation of how children's communicative interactions via gesture become codified into the initial I-language elements of the new language. If it is correct that this is a common experience for deaf sign users, this work will likely lead to an outpouring of comparable data from different contexts, heralding new insights into the emergence of new languages.

If successful language acquisition may be triggered even by exposure to very restricted data, then perhaps children learn only from simple expressions. They only need to hear simple expressions, because there is nothing new to be learned from complex ones. This is degree-zero learnability, which hypothesizes that children need access only to unembedded material (Lightfoot 1989). Such a restriction would explain why many languages manifest computational operations in simple, unembedded clauses but not in embedded clauses (e.g., English subject-inversion sentences like *Has Kim visited Washington?* but not comparable embedded clauses *I wonder whether has Kim visited Washington*), whereas no language manifests the reverse: operations that appear only in embedded clauses and not in matrix clauses.[4] One explanation for this striking asymmetry is that children do not learn from embedded domains. Therefore, much that children hear has no consequences for the developing I-language; nothing complex triggers any aspect of I-languages.

In 2014, Norbert Hornstein and Bill Idsardi, working with a pre-publication version of Heinz 2016, proposed that investigating whether children's I-languages are influenced only by simple structures, specifically from unembedded binding domains, be taken as a central "Hilbert problem" for linguistics. That is, a problem or area of investigation where advances would influence the future of the field. Postulating degree-zero learnability has already led to productive reanalyses of phenomena that *seemed* to suggest that children learn from embedded material; there are now significantly better analyses that are based on simple triggers (Lightfoot 2012). That kind of analytical productivity is the hallmark of a Hilbert problem. For related discussion, see also Heinz and Idsardi 2011 and 2013. This Hilbert problem also goes beyond approaches to language acquisition that view it as a process of setting binary parameters.

Before we turn our attention to variable properties in the next section, let us first consider Charles Yang's Tolerance Principle, a general invariant principle he has recently discovered that governs irregularity. Yang 2016 shows that children make a categorical distinction between rules and their exceptions and offers a computational account that relates patterns of regularity to the number of possible irregular forms. Yang draws on thinking in classical economics about the price of goods reflecting a balance between supply and demand. In this view, children discover a productive rule only if it yields a more efficient organization of language, with the number of irregular forms falling below a well-defined threshold. Yang argues for his principle by showing what it predicts for irregular linguistic phenomena across countless languages. A striking property of the principle is that it makes predictions about where generalizations may also be formulated for nonlinguistic cognitive domains. There must be a principled distinction between a core I-language and peripheral irregularity. Given Yang's Tolerance Principle, for a linguistic rule to be productive, the number of exceptions must fall below a critical threshold that can be calculated precisely.

Recall that this chapter is entitled "Three Visions." Our first vision is that of Minimalists who have focused on invariant principles. That focus has been remarkably successful in generating understanding of what happens in that astonishing third year of life when a child matures into a full-fledged member of the human species. However, while Minimalists are well aware of the interrelated problems of language variation and language acquisition, they have had strikingly little to say about them.

Obviously, there is much more to say about our first general vision. Indeed, Chomsky's Uniformity Condition, to be discussed in the next section, seems to indicate that there is no variation at the level of I-languages and that all children attain one, invariant language, Human, stated at a level of abstraction that embraces a competence in English, French, Javanese, Japanese, or any other of the apparently diverse languages. We will return to the matter of Human and an abstract skeletal structure again in §1.3 and in chapter 6. We will eventually argue for a form of that vision, by arguing that UG is "open" and that variable properties are learned through the parsing operations. Meanwhile, let us consider the difficulties generativists have had in characterizing variable properties in I-languages and their acquisition, specifically the problems with parameters and how they are set.

1.2   Parameters and Their Problems

Postulating hierarchical linguistic structures formed by a simple Merge operation has yielded new understanding of the invariant properties of language and has generated an immensely fruitful research program, bringing explanatory depth to a wide range of phenomena (Den Dikken 2012). However, a hallmark of human language, alongside its invariant properties, is its apparent variation. Some properties occur only in certain I-languages, not in others; these are the variable properties. The second vision motivating some linguists focuses on those variable properties. As Charles Yang (2016: 1) puts it,

A theory of language needs to be sufficiently elastic to account for the complex patterns in the world's languages but at the same time sufficiently restrictive so as to guide children toward successful language acquisition in a few short years (Chomsky 1965).

The environmentally induced variation that one finds in language is biologically unusual, not what one sees generally in other species' faculties or in other areas of human cognition, and requires a biologically coherent treatment. Children appear to attain significantly different internal languages, depending on whether they are raised in contexts using some form of Swedish or a kind of Vietnamese. English speakers in seventeenth-century London typically acquired different grammars from those acquired three generations earlier. Furthermore, people speak differently depending on their class background, their geography, their interlocutors, their mood, their alcohol consumption, and other factors.

For a good biological understanding, variation in grammars, I-languages, needs to take its place among other types of variation. This is an area where we have made much less progress than with invariant properties and where there needs to be new thinking. Indeed, syntacticians have very different ideas about what biology provides with respect to variable properties. Although our second vision focuses on issues of variation and acquisition, its implementation has turned out not to be as successful as was hoped.

Chomsky 1981b initiated the "Principles and Parameters" approach, seeking to find a UG with both invariant principles and a set of formal parameters that children were thought to set on exposure to PRIMARY LINGUISTIC DATA. PLD are the simple data that all children can be expected to hear and that constitute the triggering experience for language acquisition. PLD do not include nonprimary data about what does *not* occur, how things might be translated, what are paraphrase relationships or the scope of quantifiers, or other exotic data that are useful for linguists figuring out the best analysis for a language's syntax but are largely irrelevant to children's acquisition.

For four decades, linguists have been postulating parameters, ideally binary parameters (either structure *a* or structure *b*, for example, head-final or head-initial), but no real, general theory has emerged, and genuinely binary parameters are scarce. The key idea is that children set parameters depending on what they experience (deciding whether their PLD require a head-initial parameter setting, for example). Minimalists set on reducing the complexities of the invariant principles that had emerged by the mid-nineties have not devoted equivalent efforts to minimizing the complexities of UG parameters, the variable properties, nor, more importantly, to giving an account of how parameter settings might be acquired by young children. Scientists following the parameter-based vision have sought a theory of variable properties by defining those properties at the level of UG, which incorporates a set of parametric options alongside the invariant principles. A much cited metaphor equates parameter settings to on–off switches that children flip in response to the PLD that they experience.[5] The vision was that those switches would have multiple consequences, capturing "harmonic" properties. For example, a language with null subjects, like Italian and Spanish, would also manifest complex verbal morphology, *that*–trace violations, and other related properties.

The term *parameter* was first introduced into the syntactic literature by Luigi Rizzi in footnote 25 of Rizzi 1978, when he "parameterized" bounding nodes for the Subjacency Condition. Different I-languages have different phrasal categories serving as bounding nodes, and Rizzi addressed acquisition issues by postulating markedness relations between them. Elaborating the Principles and Parameters approach to universal and variable properties, Chomsky 1981b takes principles to embody the universal properties of I-languages and parameters to embody the limits on variation in I-languages. Parameters have been used for quite different forms of variation: sometimes for specific structural properties, such as whether IP might count as a node limiting long-distance movement (Rizzi 1978); sometimes defining language "types," such as "null-subject languages," which have null subjects and related harmonic properties (Roberts & Holmberg 2010; and see Haider 2005 for a "typological" approach to differences between Icelandic and German and Cinque 2013 for broader considerations); sometimes distinguishing parameters and microparameters for different scales of variation (Baker 2008; Kayne 1996; Westergaard 2009, 2014, 2017), perhaps keying microvariation to features on functional categories as in Kayne 2005; or even keying *all* variable properties to features on functional categories, as in the Chomsky–Borer Conjecture (Borer 1984; Chomsky 1995). A fundamental observation that motivated the idea of parameters was that variation does not come in random ways but in *clusters*. Parameters, whatever they are exactly, are abstract and designed to capture the clusters of phenomena that make up actual points of variation between I-languages. If we argue against the existence of binary parameters, we will need another way of capturing the harmonic properties that variation manifests, a matter that we return to in chapter 5.

Sometimes researchers invoke what is called the Chomsky–Borer Conjecture as a theory of parameters, saying that parameters constitute features on functional categories. Restricting variation to elements of morphology (lexical features of functional categories) avoids any direct, clear role for structures and leaves much unsaid; it cannot be a restrictive theory of parameters until we have a substantive theory of lexical features, and it pushes matters back to asking what are the lexical features and what theoretical ideas might unify them.[6]

The vision behind this conception of on–off parameter switches is that points of variation fall into a narrow class, with perhaps just thirty or forty parameters defined at UG. Under this second of our three visions, variation would be expected to be narrowly defined and limited. However, when one looks at the kind of variation attested in I-languages, the vision of a limited number of points of structural variation does not look plausible; one would expect the variation to be less varied than it in fact appears to be. For example, the parade case of a parameter, the null-subject parameter, takes many different forms, and one sees languages developing very narrow and specific new variable properties (chapters 2 and 3). Even so, the parameter-based vision presents an important challenge for Minimalist thinking, since it attributes a great deal of information to the linguistic genotype, the kind of rich information that has proven dispensable in the UG-defined invariant principles. However, there are other problems with this conception of grammatical variation, which suggest that it may be time to try another approach (cf. also Boeckx 2014; Epstein, Obata, & Seeley 2017).

Meanwhile, there is no coherent definition of parameters and the term is used to refer to quite different things, as just noted. And separately, there has been very little discussion of how children set the parameters that have been proposed. Indeed, variation in parameter settings does not have the explanatory depth that we see in invariant principles of UG, and I am not aware of arguments for any particular parameter-based form of poverty-of-stimulus reasoning. But Crain 2012 does argue that one needs parameters to understand the scope properties that children demonstrate when acquiring English and Mandarin quantifiers; I have not yet worked on these cases to see how a parsing approach might fare. Parameters used to be the only way of approaching issues of variability in mainstream generative work, but now we have parsing-based approaches. For example, Getz 2018a tackles an old problem treated in terms of parameters and, developing an analysis based on parsing, shows how children can learn the relevant distinctions. Newmeyer 2004 holds that parameters have not illuminated the nature of linguistic variation and that linguists have been unprincipled in their use of them (but for critical discussion see Biberauer 2008, including the editor's introduction, and Holmberg 2010 and Roberts & Holmberg 2010).

Linguists study variation in silos: syntacticians studying parameters have little contact with sociolinguists studying variable rules, and proponents of variable rules do not interact much with Optimality theorists studying constraint reranking, who, in turn, do not collaborate with cartographers seeking the variable properties associated with different functional heads. Indeed, Minimalists have devoted little attention to variation and acquisition (the two go together: variable properties must be acquired by children during development, whereas invariant properties may be provided in advance by UG and need not be triggered by primary data). The idea of UG parameters (macro- and micro-) grossly violates Minimalist aspirations to minimize information at UG. Indeed, Chomsky 2001 invoked a Uniformity Principle, abstracting away from variation: "in the absence of compelling evidence to the contrary, assume languages to be uniform, *with variety restricted to easily detected properties of utterances*" (my emphasis). This opens the logical possibility that all children attain the same I-language, which might be called Human, whether they are raised in a Javanese- or Japanese-speaking environment. Human would be stated at a level of abstraction that would embrace the specificities of Javanese or Japanese. Chomsky has never argued for such a possibility, as far as I know, but it would comport with his Uniformity Principle, and we will in §1.3 suggest a way of implementing this idea in terms of UG being open to information that is learned through parsing.

Hornstein's 2009 effort to transform the Minimalist Program into a Minimalist Theory has essentially no discussion of variation or acquisition, apart from a brief, four-page discussion of the history of parameters (pp. 164–168). Boeckx 2015 goes further and seeks to eliminate lexically determined features on the remarkable grounds that "they are obstacles in any interdisciplinary investigation concerning the nature of language," as if linguists should deal only with analytical machinery invoked by present-day biologists.[7]

Difficulties with parameters are aggravated by the absence of an adequate account of which PLD set which parameters and how. It is often supposed that children evaluate candidate grammars by checking their generative capacity against a global corpus of data experienced, converging on the grammar that best generates all the data stored in the memory bank. But that entails elaborate calculations by children and huge feasibility problems (Lightfoot 2006a: §4.1).

The best-worked-out evaluation systems are Robin Clark's FIT-NESS METRIC (1992) and Gibson and Wexler's TRIGGERING LEARN-ING ALGORITHM (1994). Both systems involve the *global* evaluation of grammars as wholes, where the grammar *as a whole* is graded for its efficiency (Yang 2002). Children evaluate postulated grammars as wholes against the whole corpus of PLD that they encounter, checking which grammars generate which data. This is "input matching" (Lightfoot 1999) and raises questions about how memory can store and make available to children at one time everything that has been heard over a period of a few years.

Clark's genetic algorithm employs his Fitness Metric to assign a precise number to each grammar based on what it can generate and how many "violations" there are, that is, how many unacceptable structures the grammar generates wrongly. The key idea is that certain grammars yield an understanding of certain sentences and not others; put differently, they generate certain sentences and not others. The Fitness Metric quantifies the failure of grammars to parse sentences by again counting the "violations," in this case, sentences experienced that cannot be generated by the grammar being evaluated. There are two other factors involved in his Fitness equation, a superset penalty and an elegance measure, but those factors are subject to a scaling condition and play a minor role, which I ignore here. The Fitness Metric remains the most sophisticated and fully worked-out evaluation measure that I know. It is a global and very precise measure of success, assigning specific indices to whole, fully formed grammars against a whole corpus of sentences, rating their success exactly according to the extent that their output matches the child's input.[8]

Gibson and Wexler take a different approach, but in their view as well, children effectively evaluate whole grammars against whole sets of sentences, although they react to local particularities. They are "error-driven," and they determine whether a certain grammar will generate everything that has been heard and whether the grammar matches the input. Gibson and Wexler's children acquire a mature grammar eventually by using a hypothetical grammar (a collection of parameter settings) and revising it when they encounter a sentence that their current grammar cannot generate (an "error" in Gibson & Wexler's terminology, corresponding to Clark's "violation"). In that event, children follow Gibson and Wexler's Triggering Learning Algorithm to pick another parameter setting, and they continue until they converge on a grammar for which there are no unparsable PLD and no errors. Gibson and Wexler 1994 uses a toy system of three binary parameters that define eight possible grammars, each of which generates a set of sentence types. The child essentially tests the generative capacity of grammars in light of the complete set of sentence types experienced. Both Clark's child and Gibson and Wexler's child *calculate* the degree to which the generative capacity of the grammar under study conforms to what they have heard.

Even considering a small number of parameters, the problems become clear. If parameters are independent of each other, forty binary parameters entail over a trillion possible grammars, each generating an infinite number of structures. Parameters, of course, are not always independent of each other, and therefore the number of grammars to be evaluated might be somewhat smaller. On the other hand, Longobardi et al. 2013 postulates fifty-six binary parameters just for the analysis of noun phrases in Indo-European languages, which would suggest much larger numbers. On any account, the relevant numbers are astronomical.

It is worth bearing in mind that all I-language variable properties result from change (assuming that all languages have a common, single, monophyletic origin). Alongside addressing feasibility issues, one's theory of variation must accommodate grammatical change from one generation to another. Seeking to explain changes through language acquisition requires an approach to language acquisition different from input matching. If children acquire grammars by evaluating them against a corpus of data, then for an innovative grammar, they need to be confronted with the data generated by this grammar in order to select the grammar that generates them. This introduces problems of circularity: what comes first, the new grammar to generate the new data or new data that require the child to select the new grammar? Given that change seems to be ubiquitous and is the source of all variation, this circularity is a reason to discount the input-matching approach to the acquisition of variable properties.

There are other conceptual problems with viewing children as evaluating the generative capacity of numerous grammars, calculating how best to match the input experienced, and setting binary parameters accordingly (see also Boeckx 2006, 2014; Lightfoot 2006a; Newmeyer 2017). Certainly, given the abstractness of grammars, it will not do to suppose that triggers for elements of I-languages are sentences that those elements serve to generate, as seems to be assumed by Crain and Thornton (1998); that would introduce other problems of circularity.[9]

To illustrate the lack of explanatory power in parameters, consider a formal parameter, postulated to be part of UG, according to which a head either precedes or follows its complement: {head,complement}. An alternative is to say that a child parses head-first expressions like [$_V$sing $_{DP}$[three songs]] and [$_N$books $_{PP}$[about cities in California]] as such. The "ordering parameter" solves no poverty-of-stimulus problem and does not reduce the information needed for a child to converge on the right structure. Similarly, it is not helpful to say that languages like Dutch have a parameter set to yield verb-second structures like $_{DP}$[drie studenten] $_{CP}$[$_C$bezoeken $_{VP}$Utrecht] 'three students visit Utrecht', as opposed to saying that children parse such structures and consequently have I-languages that generate such structures. It would also not be explanatory to say that English I-languages have a parameter set to allow complex DPs like *the man from Utrecht's daughter with long hair* as opposed to saying that children parse such expressions with *'s* analyzed as a Determiner clitic licensing the complex DP *the man from Utrecht*: $_{DP}$[the man from Utrecht] $_{Det}$'s $_{NP}$[daughter with long hair]. Rather, UG is open to such structures being postulated when children are exposed to relevant ambient language that requires or EXPRESSES such structures.

In summary, combining our first two visions so that I-languages are viewed as consisting of invariant properties and a set of formal parameter settings, with both principles and parameters being defined at UG, fails to account for language acquisition and language variation. In particular, the input-matching approach has not been successful and faces apparently insuperable difficulties. Parameters stated as part of UG have no useful role to play, and it may be opportune to consider other approaches, based on a different vision, where parsing plays a more central role.

1.3   Principled Parsing for Parameters

For our third vision, we build on earlier work that treats children as predisposed to assign linguistic structures to what they hear, as born to parse. Rather than calculating and ranking the generative capacity of grammars and setting formal parameters, children are endowed with the tools provided by a restrictive and Minimalist UG that has no specific parsing procedures (Lightfoot 2017b). They assign structures to what they hear and use the structures necessary to interpret what is heard. Children thereby discover and select the structures, in principle one by one, as those structures are required by what they are hearing, as they are demanded by new parses.[10] Those structures are part of the child's emerging I-language and, in aggregate, constitute the mature I-language. Such an approach enables us to understand how children develop their internal system and how those systems may change from one generation to another, as revealed by work on historical change in syntactic systems. After all, *all* variable properties of I-languages must originate in change (Lightfoot 2018).

So internal languages consist of invariant properties over a certain domain *plus* supplementary structures that are not invariant but are required in order to parse what children hear, consistent with the invariant properties. Children select what they need to accommodate new parses, drawing on what UG makes available, notably structures built by Merge and Project as sketched in §1.1. Put differently, children parse the external language they hear, assigning to expressions structures provided by the bottom-up procedures of Project and Merge. They discover and select specific I-language elements required for certain aspects of the parse, and they do so by using what UG makes available and what they have in their emerging I-language. I-languages grow over time through parsing, since parsing enables the child to discover new contrasts in the ambient E-language, hence variable properties in the emerging I-language.

The aggregation of those parsed elements constitutes the complete I-language beyond what is given by UG.

When E-language shifts, children may parse differently, and thus a new I-language emerges. Children discover variable properties of their I-languages through parsing with the available hierarchical structures; there is no evaluation of I-languages, there are no binary parameters provided by UG, and there are no special parsing procedures. That constitutes a major minimization of current theories of grammar, declaring substantial aspects of them redundant and not needed.

There is an interplay between E-language, which is parsed, and I-languages, which result from parsing. Parsing is a direct function of an emerging I-language, not of independent parsing procedures designed to produce the structures required (more on this in chapter 3). Children discover the structural contrasts in their ambient, external language and select the structures of their emerging internal language to accommodate the contrasts.

This approach grows out of earlier work on "cue-based acquisition," which also took parsing to be a key element of acquisition (Dresher 1999; Fodor 1998a; Lightfoot 1999; Lightfoot & Westergaard 2007; Sakas & Fodor 2001). However, that early work postulated that the cues that children might discover were specified in UG, as in a restaurant menu, thereby opening itself to the objections to richly specified parameters discussed in §1.2 and predicting that new variable properties fall into a narrow class, depending on the number of parameters. Ideas that involve rich and specific information about variable properties stipulated at UG risk being biologically implausible and at variance with the aspirations of the Minimalist Program. In the current vision, variable structures, those showing up in some but not all I-languages, are not stipulated at UG but are discovered as children parse the ambient E-language, hence learned. The approach allows good understanding of why

English has adopted so many structural innovations, apparently quite idiosyncratic and unprincipled and not shared by its closest historical relatives. A researcher following this vision would expect a greater range of variation in the world's languages and would seek new, more biologically plausible ideas about variation (see chapters 5 and 6).

Two well-studied English examples very briefly; for details see Lightfoot 2017b and sections §2.4 and §3.3 here.

First, in the early sixteenth century a change was completed whereby, after the loss of very rich verbal morphology, a set of preterite-present verbs, verbs with past-tense forms but present-time meaning, came to be parsed as Inflection elements with a very different syntax from verbs, unlike in any other European language. This change has been studied by many diachronic syntacticians and is now well understood. After the widespread simplification of inflectional morphology in Middle English, only two verbal inflections survived, the third-person singular marker in *-eth* or *-s* and the second-person marker in *-st*. The preterite-presents never had the third-person marker. That absence was just one of many facts about morphology, but after the great simplification, the absence of those forms came to make these verbs categorially distinct and unlike any others. In addition, the past-tense forms of these so-called Modals like *could, would, should,* and *might,* only rarely conveyed past time, unlike *opened, thought,* and so on. Rather than expressing past time, they expressed a modal "subjunctive" meaning. The evidence is that these forms were parsed as a distinct, non-verb category of Inflection, which entailed a different syntax and the simultaneous obsolescence of formerly robust forms like *He has could see stars, She wanted to can see stars, She will can see stars,* and so on.

Because of these morphological changes, Middle English children began to discover new contrasts: the verbs *can, may, must* now ceased to look like verbs like *open, read, think,* because they lacked the third-person marker. Establishing such contrasts is the basis of parsing, and Elan Dresher's work becomes important here, seeking a general approach to contrast hierarchies in phonology and revising Roman Jakobson's work on the topic (Dresher 2009, 2019, and Cowper & Hall 2019, have extended these ideas to syntax).

Second, with the loss of case morphology on nouns, which therefore no longer occurs in the ambient E-language, a set of forty or so verbs indicating psychological states underwent a kind of reversal of meaning and a new syntax, *like* changing from 'please' to 'enjoy', *repent* changing from 'cause sorrow' to 'feel sorrow', and a theme subject changing to a patient with both verbs. With the loss of morphology, children came to parse expressions differently, assigning different structures: *Gode ne licode na heora geleafleast* 'to-God did not like their faithlessness' was no longer parsed with *Gode* 'God' as a dative case, *heora geleafleast* 'their faithlessness' as a nominative subject, and *licode* (*liked*) meaning 'pleased' or 'caused pleasure'. Instead, based on these new parses, children selected new I-languages that entailed the new structures and new semantics of these psych verbs ('God' as a nominative, 'faithlessness' as a theme, and *like* with a new meaning, 'enjoy' or 'experience pleasure'). It is hard to see how an explanation for these changes could be provided if children were evaluating multiple grammars and setting formal parameters, and it is equally hard to see what binary structural parameters might be implicated; we return to this in §3.3.

This is too brief an account, but it is time for an alternative to parameter setting as an approach to variable properties; we need something that matches our success with invariant principles. I suggest that, rather than being parameterized, UG is OPEN, consistent with a variety of structures, depending on what contrasts the parsing process reveals. Given what they experience in their ambient external language, children may or may not select nominal case endings or verbal suffixes in their I-languages; that is determined by the interplay between E-language and I-languages, and UG does not include particular parameters.

Under this approach we do not overtheorize variation by trying to specify particular options in a very rich UG. Rather, UG is unspecified in certain ways, open to demands that emerge from parsing E-language, which are accommodated by the categories, structures, and operations made available by UG and by the elements in a child's emerging I-language. Variation is inherent to the language faculty by virtue of its openness. This approach recalls the 1970s distinction between core and peripheral properties: UG provides skeletal structures that are enhanced by the results of parsing. Extending the metaphor, UG provides a skeletal structure, fleshed out by the consequences of children parsing the E-language they are exposed to. Under the vision sketched in this section, children discover variable properties by parsing the ambient E-language with the tools provided initially by UG. Through parsing, they also postulate structures in their emerging I-language that are required to understand the language(s) around them. Those variable properties are not stipulated in UG in the form of parameters to be set; that would be too much information for the stripped-down, minimized UG furnished by Minimalist thinking. This vision allows for a greater range of variable properties than thirty or forty UG-defined parameters would permit and requires a larger role for learning.

It remains to be seen whether this line of thinking yields a productive approach to variation beyond what we have viewed in the past as syntactic parameters. Hopefully, as linguists focus on the nature and acquisition of variable properties in all domains of language (see Lightfoot & Havenhill 2019), they will facilitate new approaches to variation, leading to an understanding of why human language, unlike other aspects of cognition, encompasses so much variation.

1.4 Reflections

Most discussion of variable properties concerns either properties that vary across languages or properties that vary within the PLD that an individual learner is exposed to. The former are mostly considered by generative grammarians; the latter are the focus of much work in sociolinguistic, Labovian traditions. Each of these silos could benefit from new thinking about the sources of variation, and perhaps there will be some coming together, as people see interactions between the two paradigms. After all, variation within external language triggers different individual, internal languages.

Another theme in discussion of variable properties concerns the role of statistics with respect to I-languages: statistics can complement theories of language, not replace them, if we see statistical variation within a language as something to be explained, not the explanation. Again Yang's work (Yang 2002) treats statistical information as an aid to children acquiring I-languages; statistics *reflect* properties of I-languages (see Lidz 2010; Lidz & Gagliardi 2015). And Getz 2019 shows that the age at which children learn categorical rules may restrict their nature. One can hope that different types of variation may be understood in a shared perspective and silos may become more porous.

It is interesting to reflect on how thinking has changed about the invariant principles of UG and the variable parameter settings since the early days of the Principles and Parameters approach, the early 1980s (Chomsky 1981b). In those early days the general goal was to find principles and parameters of UG that could solve the poverty-of-stimulus problems that were identified. Since researchers have sought to get "beyond explanatory adequacy," the strategy has been to reduce UG to a minimum. This does not mean that we should retreat from solving poverty-of-stimulus problems and rely on biologists to solve our analytical problems (contra Boeckx 2015; see

Lightfoot 2017b for discussion); it does mean that we need richer notions of learning beyond simply invoking properties of UG.

Two examples where this has happened: first, Getz (2018a,b) has shown that we can get better understandings of some well-known poverty-of-stimulus solutions with a richer understanding of the learning involved and how it may come from parsing ambient language. Second, Dresher, Cowper, and Hall's contrastive theory (Dresher 2009, 2019; Cowper & Hall 2019) does not assume innate features, unlike in earlier generative phonology. They can give up innate features because they postulate something else innate, contrastive hierarchies and the concept of features.

I hope that this book will help the field get to a better, more unified understanding of variable properties, in particular by recognizing that understanding their acquisition better will yield a better understanding of their nature.

Here I have explored three visions, placing bets on the third. Let us now explore that vision more fully and see what its mechanisms might offer people seeking to understand the third year of a child's life.[11] I will now sketch three new parses, three instances of new structures being discovered and selected. Meanwhile I will discard the second vision and develop the first and third.

2 Three New Parses

2.1 That Third Vision

Time now to begin to flesh out the third vision of the previous chapter and to unpack statements there to the effect that all variable I-language properties come from historical change in those I-languages. In this chapter we will reconsider three changes in the I-languages of speakers of English, two of which were addressed in my 1991 effort to show how cue-based children might set parameters (Lightfoot 1991). I addressed the third change in my 1999 book on the development of language, again treating change as a result of parametric shifts. The third vision of chapter 1 takes a very different approach, which we will now begin to work through in detail: we will view children not as setting parameters and evaluating the resulting grammars, counting what grammars do and do not generate, but instead as discovering contrasts and selecting their I-language, postulating structures as allowed by UG and as needed to understand their ambient E-language, their PLD. That is to say, children parse the external language around them, assigning linguistic structures to what they hear from that ambient E-language. Parsing the ambient language is how children discover the variable properties of their emerging I-languages.

I have asserted that any I-language-specific property must have arisen as a result of historical change. I argue that, at least in the domain of syntax, this can happen only through language acquisition by young children. Phonology might be different, but that would be another book, written by somebody else. Here I examine three very specific peculiarities of English, properties that have emerged in English I-languages and not in other European languages. We know enough about when and how they arose to be able to provide good explanations for them in terms of new parses. I will explain what changed and why. Given that very few of the world's languages have a richly recorded history, we will not be able to develop equivalent explanations in a pervasive way, but it will be helpful to think through what it would have taken for I-language peculiarities to have arisen at some point in history. A key to the whole approach is to link changes in E-language with changes in I-languages and to link them by making parsing central to our view of acquisition. As noted earlier, E-language is what is parsed and an I-language results from the parsing, hence there is an intimate relationship between the two.

2.2   Explaining Change through Learnability and Acquisition

In work on syntactic change, one needs good hypotheses for the early stage of the language under investigation and for the late stage, after the change has taken place; one needs a good synchronic analysis both before and after the changes to be considered. That means that one needs all the ideas marshaled by synchronic syntacticians. However, to describe and explain changes through time, questions arise that are not typically explored in synchronic work. Different research strategies are called for, and certainly there are different research traditions involved. Under an approach linking syntactic change to acquisition, work on change casts light on the idealizations used in synchronic work, as we shall begin to see in §2.4, and is instructive for synchronic syntacticians.

Over recent decades, an approach has developed that links explanation of syntactic changes to ideas about language acquisition, learnability, and the (synchronic) theory of grammar. One way of making this linkage construes change as always externally driven by new PLD.

Baker and McCarthy 1981 identified the "logical problem of language acquisition" as that of identifying the three elements of the following analytical triplet.

(1) Primary linguistic data (Universal Grammar → grammar)

Children are exposed to PLD and as a result their initial state, characterized as UG, develops into a mature state, characterized by a particular, individual grammar, an I-language. The solution to the logical problem lies in identifying the three items in a way that links a particular set of PLD to a particular grammar, given particular ideas about UG. Children seek the simplest and most conservative grammar compatible with both UG and the PLD that they encounter (Snyder 2007).

Under that approach, there can only be one way to explain the emergence of a new grammar. When children are exposed to new PLD that cannot be parsed appropriately by an existing I-language, the new PLD trigger a new grammar. In that sense, it is new PLD that *cause* the change; UG certainly does not cause changes, only providing the outer limits to what kinds of new I-languages may arise. PLD consist of structurally simple things (degree-zero simple, in fact; see §1.1) that children hear frequently, robust elements of their E-language.

Crucial to this approach is Chomsky 1986's distinction between I- and E-language, both of which play an essential role in explaining change. E-language is the amorphous mass of language out in the world, the things that people hear. There is no system to

E-language; it reflects the output of the I-language systems of many speakers under many different conditions, modulated by the production mechanisms that yield actual expressions. I-languages, on the other hand, are mental *systems* that have grown in the brains of individuals who have parsed their ambient E-language. These systems characterize the linguistic range of those individuals; I-languages are represented in individual brains and are, by hypothesis, biological entities.

New PLD cause change in I-languages. Work in diachrony, therefore, makes crucial use of the E-language–I-language distinction and keys grammatical properties to particular elements of the available PLD in ways that one sees very rarely in work on synchronic syntax. Successful diachronic work distinguishes and links two kinds of changes: changes in PLD (part of E-language) and changes in mature I-languages. These changes are quite different in character, as we will see when we consider our well-understood changes below (§2.4–§2.6) and ask, more broadly, about the role of PLD and whether children are setting parameters or discovering new structures.

Work explaining language change through acquisition by children has been conducted now for decades, and there have been surprising results that lead us to rethink the relationship between PLD and particular I-languages. Diachronic syntacticians have ideas quite different from those common among their synchronic colleagues about which PLD trigger which particular grammars, something that synchronic syntacticians rarely write about.

2.3   Models of Acquisition

Work in synchronic syntax has rarely linked grammatical properties to particular triggering effects, in part because practitioners often resort to a model of language acquisition that is flawed and strikingly disconnected from work on historical change. I refer to a model that sees children as evaluating grammars globally against sets of sentences and structures, matching input, and evaluating grammars in terms of their overall success in generating the input data most economically (e.g., Clark 1992; Gibson & Wexler 1994, discussed in §1.2). A fundamental problem with this approach is that I-languages generate an infinite number of structures. If children are viewed as setting binary parameters, they must, on the conservative assumption that there are only thirty or forty parameters, entertain and evaluate billions or trillions of grammars, each capable of generating an infinite number of structures (see §1.2).

Beyond these overwhelming issues of feasibility, the evaluation approach raises further problems for thinking about syntactic change, because work often fails to distinguish E-language changes from I-language changes and encounters problems of circularity: the new grammar is most successful in generating the structures of the new system, but in order to explain the emergence of the new grammar, it is presupposed that the new structures are already available. This is part of a larger problem: if one asks a syntactician how children can learn some grammatical property, she will point to sentences that are generated in part through the effects of the relevant grammatical property, taking those sentences to be the necessary PLD. This circularity will become clearer when we discuss specific changes.

New thinking is needed. A "discovery approach" is not subject to the feasibility problems of global grammar evaluation if it treats children as selecting structures expressed by PLD, the structures needed in order to parse sentences. There may be a thousand or more possible structures, but that does not present the feasibility problems of evaluating the success of thirty or forty binary parameter settings against corpora, that is, evaluating the comparative success of complete I-languages with different parameter settings in generating a given corpus of structures. Children posit structures that are required to analyze what they hear, and that parsing is the key to language acquisition (as suggested by Janet Fodor's important pair of papers: Fodor 1998b,c). Once children have an appropriate set of structures, the resulting I-language generates what it generates and the overall set of structures generated plays no role in triggering or selecting the grammar. That is, children do not perform calculations on what different grammars generate. Children parse, selecting the structures needed to understand what they hear, in principle, one by one in local decisions. A particular grammar is the result, but children do not evaluate the generative capacity of different overall grammars.

This model of acquisition, essentially a discovery procedure in the sense of *Syntactic Structures* (Chomsky 1957; see my introduction to the second edition, Lightfoot 2002), keys elements of grammar to particular elements of the PLD and provides good explanations for diachronic shifts and the emergence of new grammars. Our discovery procedure, however, is complemented by a strong theory of UG, unlike the discovery procedures characterized in *Syntactic Structures*. Under this model, we can link an element of I-language structure with PLD that express that structure, and this yields some surprising results that force us to think about triggering experiences differently.

So a person's internal language capacity is a complex system that depends on an interaction between learned operations and principles that need not be learned but are conveyed by the genetic material, directly and indirectly. The language capacity grows in children in response to the E-language that they encounter, the source of the I-language structures, and becomes part of their mature biology. If language growth in young children is viewed in this way, then we can explain language change over generations of speakers in terms of the dynamics of these complex systems: new I-languages are driven entirely by children responding to new E-language. In particular, we explain how languages shift in bursts, in a kind of punctuated equilibrium, and we explain the changes without invoking principles of history or ideas about a general directionality or about dispreferred grammars (unlike much modern work on diachronic syntax). The three I-language innovations to be discussed in the next three sections are completely contingent on the new PLD in changing E-language.

Under this approach, there is no separate theory of change, no distinct mechanism dealing just with change. Sometimes there are changes in E-language such that children are exposed to different PLD that trigger a different I-language, as illustrated in §1.3. New I-languages, in turn, yield another new set of PLD for the next generation of children in the speech community. That new E-language, stemming in part from the new I-languages, helps to trigger another new I-language, with further consequences for E-language. If we understand indirect connections in this way, we can explain domino effects in language change (see §2.7 and chapter 5).

This chapter examines a sequence of three reanalyses in the I-languages/grammars of English speakers, three PHASE TRANSITIONS, sets of simultaneous changes that introduced unusual properties not shared by closely related or neighboring languages. In all cases, children are computationally conservative, selecting the simplest I-language consistent with principles of UG and the ambient E-language and PLD (Snyder 2007); in particular, we do not need special principles to weed out "dispreferred" I-languages.

Earlier attempts to formulate a discovery approach to language acquisition (e.g., Dresher 1999; Fodor 1998a; Lightfoot 1999) postulated a set of "cues," structures provided by UG that children discovered when encountering PLD that needed such structures. That involved postulating rich information at the level of UG, very much against the spirit of the Minimalist Program. In this book, I present the changes quite differently, treating the relevant structures as emerging when children parse the ambient PLD.

Here I show how this works in the new parses emerging in I-languages of English speakers; in the next chapter I will focus more on the novel approach to parsing that I adopt.

2.4   First New Parse: English Modals

Modern English has forms like the (a) examples in (2–6) but not the (b) examples.

(2) a.   He has seen stars.
     b.   *He has could see stars.

(3) a.   Seeing stars, ...
     b.   *Canning see stars, ...

(4) a.   He wanted to see stars.
     b.   *He wanted to can see stars.

(5) a.   He will try to see stars.
     b.   *He will can see stars.

(6) a.   He understands music.
     b.   *He can music.

However, earlier forms of English also had the (b) forms; they occur in texts up to the writings of Sir Thomas More in the early sixteenth century. More and some writers before him used all the forms of (2–6), but the (b) forms do not occur in anybody's writing after him. In (7–9) are examples of the latest occurrences of the obsolescent forms, (7) corresponding to (2b), (8) to (4b), and (9) to (5b).

(7) If wee **had mought** convenient come togyther, ye woulde rather haue chosin to haue harde my minde of mine owne mouthe.
     (1528; More, *A Dialogue Concerning Heresies*)
     'If we had been able to come together conveniently ...'

(8) that appered at the fyrste **to mow** stande the realm in grete
stede
(1533; More, *Apology*)
'what appeared at first to be able to stand the realm in good
stead'

(9) I fear that the emperor will depart thence, before my letters
shall may come unto your grace's hands.
(1532; Cranmer, letter to King Henry VIII)

There is good reason to believe that there was a single change in
people's internal systems: *can, could, must, may, might, will, would,
shall, should,* and *do* were previously parsed as more or less normal
verbs, but they came to be parsed as Infl (Inflection) or T (Tense) elements. Before More, verbs like *can*, in fact all verbs, MOVED to a
higher Infl position, as in (10). After More, verbs like *can* were generated directly as Infl elements and occurred in structures like (11).

(10)   (11)

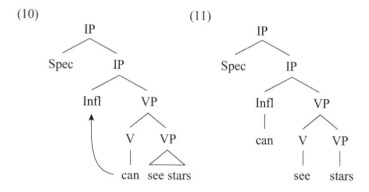

This single shift in the system was manifested by the simultaneous
loss of the (b) forms in (2–6): the phase transition. Sentences like
these are not compatible with a system with structures like (11)
instead of structures like (10). If perfect and progressive markers

are generated in the specifier of VP, then they will never occur to the left of Infl as in (2b) and (3b). If there is only one Infl in each clause, then (4b) and (5b) will not be generated; (6b) could not be generated by structures like (11), which do not allow Infl to directly precede a DP like *music*.

This change occurred only in Early Modern English and no other language; it was complete by the early sixteenth century. Notwithstanding its uniqueness, positing that this change is somehow indicative of a general tendency does enable researchers to unify it with other phenomena, which offers some degree of explanation. The change of category membership for the English modals is, in fact, a parade case of grammaticalization; but saying that it results from an internal drive, or a general tendency, or a UG bias in that direction gives no explanation for why it happened when it did and under the circumstances under which it did. Nothing similar happened in any other European language, so this change cannot be explained by a "general tendency" to grammaticalize or to recategorize modal verbs as members of a functional category.

A critical property of this change is that it consisted entirely in the *loss* of the (b) phenomena in (2–6), with no new forms emerging. Since children converge on their I-language in response to ambient simple expressions and not in response to negative data about what does not occur, the new, more limited data need to be explained by a new abstract system that fails to generate the (b) phenomena. There were no new forms in which the modal auxiliaries began to occur, so the trigger for the new system must lie elsewhere. In this case, the new PLD cannot be the new output of the new grammars, because there are no new forms. Changes like this, which consist only in the loss of expressions, make a kind of poverty-of-stimulus argument for diachrony: there appear to be no new forms in the PLD that directly trigger the loss of those expressions.

If we ask why this or any other I-language change happened, there can only be one answer under this approach: Children came to have different PLD as a result of a prior change in E-language. We have a good hypothesis about what the change was in this case.

Early English had complex inflectional morphology. For example, given the inflection of verbs for person and number, we find *fremme, fremst, fremþ, fremmaþ* in the present tense of 'do' and *fremed, fremedest, fremede, fremedon* in the past tense; *sēo, siehst, siehþ, sēoþ* in the present tense of 'see'; *rīde, rītst, rītt, rīdaþ* for the present tense of 'ride' and *rād, ride, rād, ridon* for the past tense. There was a massive loss of verbal morphology in Middle English, beginning in the north of England and due to intimate contact with Scandinavian speakers and widespread English–Norse intermarriage and bilingualism. Again I skip interesting details (see Lightfoot 2017a and §5.4 here), but external language that children heard changed such that the modern modal auxiliaries *can, shall,* and so on came to be morphologically distinct from other verbs. As members of the small preterite-present class, they lacked one surviving feature of person-and-number inflection, the present-tense third-person singular ending *-s*; they had the other surviving feature, the past and present second-person forms in *-st,* but that seems not to have been enough to ensure their survival as forms of verbs. This made them formally distinct from all other verbs, which had the *-s* ending. Furthermore, their "past-tense" forms (*could, would, might,* etc.) had meanings that were not past time, reflecting old subjunctive uses:

(12) They might/could/would leave tomorrow.

The evidence indicates that these modal verbs were parsed differently in people's internal systems, because they had become formally distinct from other verbs as a result of the radical simplification of morphology (Lightfoot 1999). Thomas More parsed his E-language and had elements like $_{\text{Infl}}[_{\text{V}}can]$ in his I-language (a verb *can* moved to an Infl position), while after him speakers had $_{\text{Infl}}can$, where *can* was no longer parsed as a verb. So we see domino effects: changes in what children heard, the newly reduced verb morphology, led to a different categorization of certain verbs, which yielded systems like (11) that were compatible with the (a) forms of (2–6) but not with the (b) forms.

Thomas More is the last known person with the old system. For a period, both systems coexisted: some speakers had (10) and others had (11), the former becoming rarer over time, the latter more numerous. A large literature is now devoted to this kind of sociological variation, changing over time, and we will return to this matter in §2.7.

Parsing, of course, depends on contrasts: in particular, formal and distributional contrasts influence the categorization of words. Words like *kick*, *like*, and *seem* are formally marked for person, number, and tense, co-occur with adverbs like *often* and *tomorrow*, and are categorized as verbs. *Girl*, *caterpillar*, and *catastrophe* are formally marked for number but not person or tense, co-occur with determiners like *the* and adjectives like *tall* and *unbelievable*, and are categorized as nouns. As a result of the simplification of morphology, verbs became definable as having the -*s* ending. In that case, *can* was no longer parsable as a verb, was categorized differently, and as a result developed a new distribution. This all makes sense, assuming uncontroversially that words are parsed as members of syntactic categories and that the parses might change over time. However, it is hard, not to say impossible, to see how the new variable properties could be seen as manifesting a new setting of a binary and UG-defined structural parameter, despite Lightfoot 1991.

2.5   Second New Parse: Verbs Ceasing to Move

A later major change was that English lost forms like the (a) forms in (13–15), another phase transition of simultaneous disappearances. Such forms occur frequently in texts up through the seventeenth century, diminishing over a long period in favor of the (b) forms and ultimately disappearing.

(13) a. *Sees Kim stars?
     b. Does Kim see stars?

(14) a. *Kim sees not stars.
     b. Kim does not see stars.

(15) a. *Kim sees always stars.
     b. Kim always sees stars.

Again we can understand the parallelism of the changes in terms of a single change in the abstract system, namely the loss of the operation moving verbs to a higher Infl position:

(16)

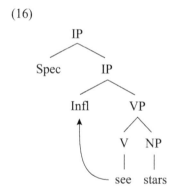

This is another change that did not affect other European languages, whose systems have retained the verb-movement operation (apart from Faroese and, perhaps, some Scandinavian systems; see Heycock et al. 2012 and references therein). Present-day English verbs do not move to the higher position and therefore cannot move to a clause-initial position (13a), to the left of a negative (14a), or to the left of an adverb (15a). The equivalent movements continue to occur in French, Italian, Spanish, Dutch, and German systems. Again a contingent explanation is required: what was it about English at this time that led to this shift in I-languages? In particular, what were the new PLD that children were exposed to that helped to trigger the new parse?

It is plausible that this shift was due to two prior changes in E-language and that we see here another domino effect, a sequence of changes. The first change was the output of the new I-language that we just discussed, involving the new parsing of a distinct category of modal verbs, which are very frequent in typical speech (see Leech 2003). Given that words like *can* and *must* were no longer verbs but Inflection items, no sentence containing such a word would have a $_{Infl}V$ structure.

The second change was the emergence of "periphrastic" *do* forms as an alternative option for expressing past tense: *John did leave, John did not leave* instead of *John left* and *John left not*. Given that *do* forms were instances of Inflection, any sentence containing one would not have the $_{Infl}V$ structure. As a result of these changes, beginning on a large scale in the fourteenth century, the Infl position came to be heavily occupied by modal auxiliaries and *do*. Thus, lexical verbs did not occur in the Infl position as often as before the days of periphrastic *do* and before modal auxiliaries were no longer verbs, and as a result the $_{Infl}V$ structure was expressed much less. Apparently it fell below the threshold that had permitted its selection by children. The loss of the (a) forms in (13–15)—a loss that reached totality only in the eighteenth century—suggests that a new system emerged in which the Infl position was no longer available as a target for verb movement. The conservative Thomas More had parsed his E-language to have $_{Infl}V$ structures, reflecting the movement of verbs to Infl; after the eighteenth century individuals had no such structures but instead had $_V[V+Infl]$, reflecting the attachment of inflectional endings onto a lower, unmoved verb (subject to Howard Lasnik's Stranded-Affix Filter, ensuring that affixes are attached somewhere: Lasnik 2000: 123).

As with the first new parse, the two systems coexisted for a while, in fact for a longer period in this case: Shakespeare and other writers alternated easily between the coexisting old and new systems, sometimes using the old V-to-Infl forms and sometimes the new *do* forms, even in adjacent sentences, as in the following examples from Shakespeare's *Othello*.

(17) a. Where **didst thou** see her?—O unhappy girl!—With the Moor, **say'st thou**?
  b. I **like not** that.—What **dost thou** say?
  c. Alas, what **does this gentleman** conceive? How **do you**, madam?

Again this is too brief an account (see Lightfoot 2017a), but it is clear that prior changes in E-language, some due to a shift in I-languages, had the effect of reducing enormously children's evidence for the $_{Infl}V$ structure, triggering a new internal system and a new parse. These simultaneous but apparently unrelated changes were a function of that single change in the abstract system, the loss of $_{Infl}V$ structures, a genuine phase transition consisting of diverse but simultaneous changes.[1]

2.6 Third New Parse: Atoms of *Be*

There is a third, remarkable phase transition, observed and analyzed in Warner 1995, which results in part from the two changes just discussed. It involves very peculiar properties of the verb *be*, which have no equivalent in other European languages that are closely related to English. It certainly lies far beyond any viable theory of parameters. One way of characterizing the change is that different forms of the verb *be* came to be listed in the mental lexicon as atomic, or "monomorphemic" as Warner puts it, and developed their own subcategorization frames.

First, some background relating to VP ellipsis. VP ellipsis is generally insensitive to morphology and one finds cases where the understood form of the missing verb differs from the form of the antecedent:

(18) a. Kim **slept** well, and Jim will [sc. **sleep** well] too.
    b. Kim **seems** well-behaved today, and she often has [sc. **seemed** well-behaved] in the past, too.
    c. Although Kim **went** to the store, Jim didn't [sc. **go** to the store].

There is a kind of SLOPPY IDENTITY at work here, since *slept* and *sleep* in (18a) are not strictly identical but "sloppily" so. Similarly in (18b) and (18c). One way of thinking of this is that in (18a) *slept* is analyzed as [$_V$*sleep* + past] and the understood verb of the second conjunct accesses the verb *sleep*, ignoring the tense element.

However, Warner noticed that *be* works differently: it occurs in elliptical constructions only on condition of STRICT IDENTITY with the antecedent. In (19a,b) the understood form is strictly identical to the antecedent, unlike in the nonoccurring (19c–e).

(19) a. Kim will **be** here, and Jim will [sc. **be** here] too.
    b. Kim has **been** here, and Jim has [sc. **been** here] too
    c. *Kim **was** here and Jim will [sc. **be** here] too.
    d. *If Kim **is** well-behaved today, then Jim probably will [sc. **be** well-behaved] tomorrow.
    e. *Kim **was** here yesterday and Jim has [sc. **been** here] today.

This suggests that *was* is not analyzed as [$_V$*be* + past], analogously to *slept*, and that forms of *be* may be used as an understood form only when precisely the same form is available as an antecedent, as in (19a,b).

Warner notes that the ellipsis facts of modern English *be* were not always so, and one finds forms like (19c–e) in earlier times. Jane Austen was one of the last writers to use such forms: she used them in her letters and in dialogue in her novels, as for example in (20a,b), but not in narrative prose, presumably indicating a conversational style. These forms also occur in eighteenth-century writings, such as in (20c), and earlier, when verbs still moved to Infl, as in (20d).

(20) a. I wish our opinions were the same. But in time they will [sc. be the same].
(1816; Jane Austen, *Emma*, ed. R. W. Chapman [London: Oxford University Press, 1933], p. 471)

b. And Lady Middleton, is she angry? I cannot suppose it possible that she should [sc. be angry].
(1811; Jane Austen, *Sense and Sensibility*, ed. R. W. Chapman [London: Oxford University Press, 1923], p. 272)

c. I think, added he, all the Charges attending it, and the Trouble you had, were defray'd by my Attorney: I ordered that they should [sc. be defrayed].
(1741; Samuel Richardson, *Pamela* [London, 3rd ed.], vol. 2, p. 129)

d. That bettre loved is noon, ne never schal.
(c1370; Chaucer, *A complaint to his lady*, line 80)
'So that no one is better loved, or ever shall [sc. be].'

These forms may be explained by supposing that, in (20a) for example, *were* is analyzed as [$_V$*be* + subjunctive] and the *be* is accessed by the understood *be* in the following *but* clause. That is, up until the early nineteenth century the finite forms of *be* are decomposable, just like ordinary verbs like *sleep* in present-day English. Hence the loss of expressions like those of (20) would be attributed to the new, monomorphemic parsing of the *be* forms.

Warner goes on to show that, now that forms of *be* are atomic and undecomposable, present-day English shows quite idiosyncratic restrictions on particular forms of *be*, which did not exist before the late eighteenth century or early nineteenth century. For example, it is only the finite forms of *be* that may be followed by a *to* infinitive, as shown by (21); only *been* may occur with a directional preposition phrase, as (22) shows; and *being* is subcategorized as not permitting an *-ing* complement, as in (23).

(21) a. Kim was to go to Paris.
     b. *Kim will be to go to Paris.

(22) a. Kim has been to Paris.
     b. *Kim was to Paris.

(23) a. I regretted Kim reading that chapter.
     b. I regretted that Kim was reading that chapter.
     c. *I regretted Kim being reading that chapter.

Restrictions of this type are stated in the lexicon. These idiosyncrasies show clearly that *been*, *being*, and so on must be listed as individual lexical entries in order to carry their own individual subcategorization restriction. However, these restrictions did not exist earlier and one finds forms corresponding to the nonoccurring sentences of (21–23) through the eighteenth century: (24a) is equivalent to (21b), (24b) to (22b), and (24c) to (23c).

(24) a. You will be to visit me in prison with a basket of provisions.
         (1814; Jane Austen, *Mansfield Park*, ed. J. Lucas [London: Oxford University Press, 1970], p. 122)
     b. I was this morning to buy silk.
         (1762; Oliver Goldsmith, *Citizen of the World*)
         Meaning: 'I went to …', not 'I had to …'
     c. Two large wax candles were also set on another table, the ladies being going to cards.
         (1726; Daniel Defoe, *The Political History of the Devil* [Oxford: Talboys, 1840], p. 336)

So there were changes in the late eighteenth century to early nineteenth century whereby the ellipsis possibilities for forms of *be* became more restricted and particular forms of *be* developed their own idiosyncratic subcategorization restrictions, both properties indicating the new, undecomposable, monomorphemic nature of the forms of *be*.

I-languages allow computational operations on items stored in a mental lexicon, and both the operations and the items stored may change over time. There is good reason to believe that decomposable items like [$_V be$ + subjunctive] and [$_V be$ + past] ceased to be stored in that form, replaced by undecomposed, atomic forms like *were*, *was*, *been*, each with its own subcategorization restrictions. The phenomenon has no parallel in closely related languages, but we can explain it by showing how it may have arisen in the I-languages of English speakers at this time.

It is natural to view this change as a consequence of the changes discussed in §2.4 and §2.5. After the loss of rich verb morphology and the loss of the $_{Infl}V$ structures, the category membership of forms of *be* became opaque, leading to new structures being assigned through new parses. If the *be* forms were instances of V, then why could they occur where verbs generally cannot occur, for example, to the left of a negative or, even higher, to the left of the subject DP (*She is not here*; *Is she happy?*)? If they were instances of Infl, then why could they occur with another Infl element such as *to* or *will* (*I want to be happy*; *She will be here*)?

In earlier English, forms of *be* had the same distribution as normal verbs. After the two phase transitions discussed earlier, they had neither the distribution of verbs nor that of Infl items. The evidence is that from the late eighteenth century, children developed I-languages that reflected new parsing, treating forms of *be* as verbs that have the unique property of moving to higher functional positions and being undecomposed, atomic elements, unlike other verbs. It is impossible to see how such specific variable properties might be captured by a binary parameter defined at UG. However, a child parsing relevant structures might indeed select the new parses that are required.

2.7   Domino Effects

Historical changes in English I-languages can be understood as the result of children acquiring their I-language as their PLD change. We have seen examples of phase transitions, when several phenomena change simultaneously. We also see domino effects, when a number of phenomena change not simultaneously but in rapid sequence. For example, English underwent massive simplification of its verb morphology, initially under conditions of bilingualism in the northeast of England (see §5.4 for more on the Scandinavian character of Middle English). The new PLD led to a new I-language with a dozen former verbs now parsed as Infl items, categorized as instances of Infl. As a result, the PLD changed again; combined with new periphrastic forms with *do*, this led to new I-languages where verbs ceased moving to higher Infl positions. This, in turn, led to new PLD in which the categorical status of forms of *be* became opaque, leading to the reanalysis of §2.6 and revealing a sequence of new parses.

The modern distinction between E-language and I-language is crucial to this analysis; both contribute to explaining change (Lightfoot 2006a). We see a dynamic interplay between changes in E-language (new PLD) and changes in I-languages. New E-language leads to new I-languages, new I-languages lead to new E-language, and sometimes we see sequences of changes, domino effects, which we can understand if a central component of language acquisition is the parsing of E-language. In the three case studies examined in the last three sections, we see causal relationships between E-language and I-language changes and the particular contingencies that triggered new I-languages at particular times.

When a new I-language, I-language$_p$, develops in one individual, that changes the ambient E-language for others, making it more likely that another child will acquire I-language$_p$; likewise for the next child, and so on. As a result, the new I-language spreads through the speech community quickly. Niyogi 2006 provides a computational model of how new language systems might spread quickly through speech communities; see §5.9.

2.8 Variable Properties

Children are exposed to speech, and their biological endowment, a kind of toolbox, enables them to begin to parse their external linguistic experience, thereby discovering I-language elements and growing a private, internal system that defines their linguistic capacity. Internal systems involve particular abstractions, categories, and operations, and these constitute the real points of variation and change; children need to discover them through parsing. Phenomena do not always change in isolation but often cluster, depending on the abstract categories involved. As a result, change is bumpy and takes place in "punctuated" bursts that disrupt general equilibrium. We explain the bumps, the clusters of changes, in terms of changes in the abstract system; the three phase transitions in the history of English that we have looked at provide an illustration. If we get the abstractions right, we explain why phenomena cluster as they do. More on this in chapter 5; meanwhile the three analyses of this chapter seem to explain the observed changes via changes in the abstract system, without needing to invoke parameters.

Everybody's experience varies and people's internal systems may vary, but not in any simple, linear fashion. I-languages change over time, and sometimes variation in E-language experience (new PLD) is sufficient to trigger the growth of a different internal system. Children are sensitive to variation in E-language, to variation in initial conditions, in the terminology of chaos theory, and this influences how they parse expressions. For example, after the comprehensive morphological changes of Middle English, young children had different experiences that led them to categorize words like *may* and *must* differently from verbs like *run* and *talk*. The assignment of these words to a different category, Infl (or T), explains why the (b) structures in (2–6) all disappeared in parallel. Similarly, new structures resulting from modal verbs being treated as Inflection items and new structures with periphrastic *do* entailed that the $_{Infl}$V structure was expressed much less robustly and fell out of use, entailing the obsolescence of the (a) versions of (13–15). Finally, as a result of these two phase transitions, forms of *be* were parsed as monomorphemic and no longer as a verb plus a separate inflectional affix.

Under this approach, change and therefore variability is contingent, dependent on particular circumstances, which accounts for why English at this time underwent changes that other European languages have not undergone at any point. English had particular morphological properties that were affected in particular ways by contact with Norse speakers and that led to the new categorization, the new parses. If change is contingent like this, then there is no general direction to change and no reason to believe that languages all tend to become more efficient, less complex, and so on. There are no general principles of history of the kind that nineteenth-century thinkers sought (e.g., Darwinians seeing "progress" in new species and Marxists seeing certain kinds of societies changing into other specific types) and that modern diachronic syntacticians continue to invoke. Explanations are local (Lightfoot 2013) and there is no reason to revive historicism, or declare principles of history, or UG biases. There is no reason to invoke UG biases or to expect that variable properties in I-languages will fall into a narrow class defined by a restrictive theory of parameters, as people following the second vision of chapter 1 imagine.

This approach to syntactic change also provides a new understanding of synchronic variation, along the lines of William Labov's 1972 discussion of sound change in progress. When a phase transition takes place, it does not happen on one day, with all speakers changing in unison. Rather, a new I-language emerges in some children and spreads through the population, sometimes over the course of a century or more but usually not for a very long period (although see the very interesting Wallenberg 2016 on the slow loss of extraposition structures). Competing grammars (Kroch 1989) explain the nature of certain variation within a speech community: in this context, one does not find random variation in the texts but oscillation between two (or more) stable organizations, two I-languages. In general, writers either have all the forms of the obsolescent I-language or none. Not all variation between texts, of course, either by the same or different writers, is to be explained in this way: only variation in I-languages.

There is also variation in E-language that has little if anything to do with I-language. E-language is amorphous, and in it variation is endemic and does not come in the structured form that variation in I-languages shows. No two people experience the same E-language, and in particular, no two children experience the same PLD. Since E-language varies so much, there are always possibilities for new I-languages to be triggered. New I-languages may come to incorporate unusual and very particular properties, as we have seen and as we shall see again when we discuss the nature of parsing in §3.1 and the spread of I-languages in §5.8. We see a wider range of variable properties than the parametric approach leads us to expect, and we understand how new variable properties emerge by seeing them as emerging through the parsing of new E-language.

2.9   Identifying Triggers

The major contributions of diachronic work in syntax lie in explaining one kind of variation, the kind due to coexisting I-languages, and in revealing, for any particular property of I-languages, what the E-language trigger might be.

It is surprising how little discussion there has been among synchronic syntacticians of what triggers what properties, given the generally accepted explanatory schema of (1), the "analytical triplet" of Baker and McCarthy 1981. Reducing hypothesis space is an essential part of the enterprise but not sufficient: we also want to know what aspects of new E-language might have triggered new I-languages.[2]

We should allow for the possibility that the PLD that trigger a particular parse may not have any obvious connection with that structure. Indeed, the recategorization of modal verbs was triggered by new morphological properties.

Niko Tinbergen (1957: chap. 22) once surprised the world of ethologists by showing that young herring gulls' behavior of pecking at their mothers' beaks was triggered not by the fact that the mother gull might be carrying food but by a red spot that mother gulls typically have under their beak. Tinbergen devised ingenious experiments showing that the red spot was the crucial triggering property. Chicks would respond to a disembodied red spot but not to a bird carrying food with the red spot hidden. Similarly and as we saw in our first case study, grammars may have a given mechanism or device as a result of properties in PLD that are not obviously related to that mechanism.

Furthermore, there is no reason to believe that elements of different I-languages are always triggered by the same PLD. For example, children developing an English I-language could learn that VPs have verb–complement order from a simple sentence like *Bent visited Oslo*. Since verbs do not move in English I-languages and

Infl lowers on to verbs, *visited Oslo* can only be analyzed as $_{VP}$[*visit* + Infl *Oslo*], verb–complement. A Norwegian child, however, could not draw the same conclusion from the word-for-word translation *Bent besøkte Oslo*, which does not reveal the underlying position of the verb. Norwegian I-languages are verb-second: finite verbs move to a high, "second" position (presumably in the C projection), yielding simple structures like $_{CP}$[*Bent* $_C$*besøkte* $_{IP}$[*Bent* $_I$*besøkte* $_{VP}$[*besøkte Oslo*]]. The verb-second analysis is required by synonymous sentences like *Oslo besøkte Bent*, where the finite verb surfaces to the left of the subject DP and requires a structure $_{CP}$[*Oslo* $_C$*besøkte* $_{IP}$[*Bent* $_{VP}$[*besøkte Oslo*]]], as well as "topicalized" expressions like $_{CP}$[*Søndager* $_C$*besøkte* $_{IP}$[*Bent* $_I$*besøkte* $_{VP}$[*besøkte Oslo*]]] 'On Sundays Bent visited Oslo'. Therefore, *Bent besøkte Oslo* does not reveal the structure of the VP, and a more complex sentence like *Bent kan besøke Oslo* 'Bent can visit Oslo' is needed to express $_{VP}$[V DP] structures, *Bent kan* $_{VP}$[*besøke Oslo*]. Similarly, in German, another verb-second language but with complement–verb order, the complex *Bent kann Oslo besuchen* reveals the $_{VP}$[DP V] structure.[3]

Speakers have their own internal system, an I-language, that grows in them in the first few years of life as a result of an interaction between genetic factors common to the species and environmental variation in PLD. Such a grammar represents the person's linguistic range, the kind of things they might say and how they may say them. If children hear different things, they may converge on a different system, perhaps the first instance of a new I-language. We want to find out what triggers which aspect of a person's I-language, to understand how new I-languages might emerge.

If we can discover things about I-languages by looking at how they change, we can generate productive hypotheses about what PLD trigger particular properties of I-languages, thereby explaining the I-language properties. That will surely interest synchronic syntacticians.

Theodosius Dobzhansky (1964: 449) noted famously that "nothing in biology makes sense except in the context of evolution," later the title of his famous 1973 paper in *American Biology Teacher*. His position was that biologists can explain properties of organisms by showing how they might have arisen through evolution. This stance is also taken by Minimalists who argue that the rich information postulated in Government and Binding approaches to UG is evolutionarily implausible. Here I have shown that there is a parallel line of argument, explaining how new language-specific properties—but not properties of UG, of course—evolve through historical change. We may have achieved ideal explanations for certain syntactic changes in terms of how children select their I-language.

2.10   Conclusion

By identifying shifts in the ambient E-language that plausibly triggered new I-languages, we can explain diachronic changes in I-languages in terms of language acquisition, distinguishing I-languages and E-language, but with each playing a crucial role. This provides a model for explaining other unusual properties that occur in mature syntactic systems. Parsing is key, implemented by the individual's I-language. Parameters and evaluation metrics play no role.

Linking matters of acquisition and learnability to matters of syntactic change permits deep explanations of particular changes and illuminates what experience it takes to trigger particular elements of I-languages. Under this approach, there is no independent theory of change, and instead change is an epiphenomenon. Children acquire their own private I-language when exposed to the ambient E-language, not influenced directly by any ambient I-language, which cannot be observed. No two children experience identical E-language. Therefore, there is always the possibility of different I-languages emerging, but nothing is actually transmitted and there is no object that changes. Rather, different I-languages emerge in different children, and I-languages with certain properties may spread through a language community, through the medium of E-language.

We explain peculiarities in I-languages by showing how they arise through language acquisition taking place when E-language changes for a generation of children. We therefore explain the peculiarities in ways that are broadly similar to the ways in which biologists explain properties of organisms by showing how they might have evolved in the species (see chapter 6). All I-language variable properties fall within a narrow range, permitted by the Merge and Project structure building allowed by the invariant principles of UG. Where we have insufficient data to tell rich stories like the ones documented here, we should nonetheless be able to imagine plausible scenarios, just as the imagination of biologists was provoked by Dobzhansky.

3 Parsing and Variable Properties

3.1 Parsing to Learn

Over a long period, several analysts of children's language acquisition have stressed the central importance of parsing: going back several decades to the early work of Bob Berwick, Janet Fodor, Mitch Marcus, Virginia Valian, and others. Parsing was the key to language acquisition for cue-based approaches twenty years ago, but we now take a strongly I-language-based approach to parsing and dispense with a parser as an independent element of cognition, simplifying and minimizing the elements of grammatical theory in the style of the Minimalists.

A central idea of this book is that, as other species are born to swim or to fly, so humans are born to parse. There is no independent parser as such; rather, parsing is implemented by a person's language faculty, initially by UG and then by an individual's emerging I-language, very differently from conventional approaches. UG provides children with the basic categories and structures built by Merge and Project (§1.1), which they can use in order to understand and parse their ambient E-language. For example, children may use words that are categorized as verbs or as prepositions and may use verb phrases structured as $_{VP}[V + Infl\ PP]$, where Infl is an Inflection element projecting to a phrasal IP node and PP is a preposition phrase containing a preposition and a determiner-phrase complement. Such an approach yields a better, more broadly based view of variable properties and superior explanations for those properties. UG is open, and variable properties emerge as children experience their E-language, discover the contrasts, and select the elements of their I-languages.

Rather than subconsciously evaluating systems against a set of data, like Clark, Gibson, and Wexler's children (Clark 1992; Gibson & Wexler 1994; see §1.2), young people as envisaged here pay no attention to what any I-language generates but instead grow an I-language by identifying and acquiring its parts (Lightfoot 2006a, 2015). Children parse the E-language they hear and discover the categories and structures allowed by UG and needed to understand what they experience, thereby accumulating the elements of their I-language. Of course, the parsing capacity matures over time and is different for a two-year-old and a three-year-old (Omaki & Lidz 2015). If there is no independent parser but if, rather, parsing is done by the developing I-language, then the parser matures as the I-language develops.

My goal here is not to flesh out a full theory of parsing but to sketch a new approach, to examine some clear cases where new parses have emerged, and to see what kinds of explanation can be offered without invoking an independent parser, with no binary, UG-defined parameters, and without any evaluation procedure, taking UG to be open and parsing to be the key to learning variable properties. After all, for us, parsing does work done by binary parameters in other models, in capturing variable properties. I shall argue in this chapter that the parameter-based vision of §1.2 fails to capture the true nature of variable properties in languages, while I-language-based parsing does a much better job of both capturing the nature of variable properties observed by comparativists and explaining how and why they arise.

Information about category membership and basic structures can be learned through parsing; robust, simple E-language triggers what is needed for an individual I-language. Our children, unlike those of Clark, Gibson, and Wexler, do not *evaluate* the generative capacity of I-languages, lining up what a system may generate and comparing that with what they have actually experienced, but discover and select the elements of their I-language through parsing the ambient E-language, building with what is provided by the toolbox of UG and required by the current I-language they are using. A child's parsing capacity is a function of UG, universal and not parameterized, and of their emerging I-language. As a child's I-language develops, the parser/I-language assigns more structures to the E-language experienced and the child interprets more E-language. Children discover variable properties of their ambient E-language, but not by setting binary parameters defined as part of UG; rather, UG is open in certain ways and, unlike in early cue-based approaches to acquisition of two decades ago, does not have predefined binary parameters.

At a certain stage of development, after they know that *cat* is a noun referring to a domestic feline and *sit* is an intransitive verb, children may experience an expression *The cat sat on the mat* and recognize that it contains a determiner phrase (DP) consisting of a determiner *the* and a noun *cat* and a verb phrase (VP) containing an inflected verb *sat* (V + Infl) followed by a preposition phrase (PP) *on the mat*, that is, $_{VP}[V + Infl \ PP]$. At this point, the child already knows the individual words expressed and the categories they belong to; that information is part of the child's emerging I-language. The child makes use of the structures needed to parse what is heard, the structures made available by UG and expressed by the E-language experienced (i.e., required for understanding); and once a structure is used, it is incorporated into the emerging I-language. In this way, a child discovers, selects, and accumulates the elements of her I-language, which are required to parse the E-language around them and to express thoughts over an infinite range; children select the elements of their I-language in piecemeal fashion and parsing changes as I-languages become richer.

At an earlier stage of development than the one just described, a logical possibility is that the child has certain words identified and some of the categories but nothing further, no phrasal categories. At that point, the child would have only a very partial structure of an E-language expression like *The cat sat on the mat*, without the phrasal units that will come later—not representing an adult's analysis of the expression, not, for example, containing the hierarchical structures that will eventually emerge in adulthood. At different points of development, a child might learn from hearing this E-language expression that there exist prepositions projecting to a PP, that *the* is a determiner projecting to a DP, that sitting on a mat is the kind of thing that a cat does.

Once children have identified several words and assigned them to their categories, they are on their way to rich syntactic structures. Minimalist theories of UG restrict the parsing possibilities quite severely: Merge puts things together into binary-branching hierarchical structures, and a verb, for example, may project to include a complement and an adjunct or specifier in the dominating VP. Furthermore, the eventual syntactic structure must meet the requirements of the two interfaces, the sensorimotor interface, formerly known as phonological form, and the conceptual-intentional interface, formerly known as logical form (see chapter 4).

People also parse complex structures that are unique to English, and with no apparent difficulty; let's tackle a really hard one. As far as I know, no European language has an exact equivalent to *each picture of Kim's*, which has a kind of quantificational meaning, specifying 'each picture out of the set of Kim's pictures'. Notice first that this expression is two-ways ambiguous: *Kim* may be interpreted as the owner or the creator of the pictures. Strikingly, however,

Kim's picture and *each of Kim's pictures* are three-ways ambiguous, allowing, in addition, a reading under which *Kim* is an OBJECTIVE GENITIVE, taken as the person portrayed. The absence of that objective-genitive reading in certain contexts needs to be explained, hopefully a direct function of the structure that the parse yields. Our expression is not one that a very young child would use, but at a certain stage our child would know that *each* is a determiner and *picture* a noun. Nouns can, in general, often be followed by a preposition phrase (*student of linguistics, road to Chicago, midfielder for Spurs,* etc.), and the complement of the preposition might itself be complex: *student of any area of science, road to any big city in Pennsylvania, midfielder for the fat man's team from Utah.* So in this case it is clear that the preposition *of* is followed by a DP, whose specifier is *Kim* and head is the determiner *'s*. Every determiner must be followed by a noun phrase, which in this example is not pronounced but "understood" to be *pictures,* and *Kim* may be interpreted as the owner or creator. *Kim* may not be interpreted as an objective genitive, however, because that would require a structure with *Kim* in the postnominal object position, from which it would have been copied and deleted: *picture of* $_{DP}$[*Kim's e ~~Kim~~*].

The deleted *Kim* (strike-through) is illicit, because a deleted item needs to be licensed by an overt, adjacent, governing head (see §1.1) and here there is only a *covert* governing head, phonologically null and not what is needed (Lightfoot 2006b). In contrast, objective genitives are permitted in forms like *Kim's picture* or *each of Kim's pictures,* which have the structure of $_{DP}$[*Kim's pictures ~~Kim~~*], with the adjacent *pictures* licensing the deletion site. We thereby explain the (un)availability of the objective-genitive readings. The expression *each picture of Kim's* is unusual, has no parallel in other languages, and tends not to be used by very young children, but it is easily parsed and understood when heard, given what the child already knows, that is, the current state of the emerging I-language. Under this analysis, the availability of objective genitives is entirely a function of the single fact that English I-languages allow *'s* to occupy determiner positions freely, where nothing similar is found in the I-languages of other European speech communities. That is the crucial difference between English I-languages and those that do not generate objective genitives like *Each picture of Kim's*.

At no stage does the child calculate what their current I-language can generate; rather they simply accumulate the necessary structures and the resulting I-language generates what it generates. Furthermore, if UG makes available a thousand possible structures for children to draw from, that raises no intractable feasibility problems comparable to those facing a child evaluating the generative capacity of the trillion grammars possible with forty parameter settings, checking these grammars against what has been heard (§1.2). It involves no elaborate calculations and requires no memory of everything that has been heard. Children developing some form of English I-language learn without apparent difficulty irregular past-tense and plural forms for a few hundred verbs and nouns. Learning that there is a structure $_{VP}[V + Infl\ PP]$ seems to be broadly a similar kind of learning, although much remains to be said. Furthermore, learning that *each picture of Kim's* lacks an objective-genitive reading poses no difficulty once we recognize that parses are subject to the constraints of the emerging language faculty.

Under this approach to parsing and to language acquisition, children do not attend to the generative capacity of their emerging I-language but attain particular elements of it step by step (see Dresher 1999 on the resulting "learning path"), and they do that by experiencing particular elements of E-language that they then parse. Work seeking to explain diachronic changes through acquisition has enabled us to link changes in E-language to changes in particular elements of I-languages, giving us a clear idea of what triggers what, as we shall see in §3.3.

3.2 Discovering and Selecting New Variable Properties through Parsing

Children may be exposed to different E-language, different PLD, than earlier members of their speech community and begin to select new I-languages that assign different structures to particular sentences. Parsing in those circumstances yields new structures. Diachronic syntacticians have identified many such innovations, often called "restructuring" or "reanalysis." However, "restructuring" is a misnomer: children grow their I-languages when exposed to ambient E-language and quite independently of the I-languages of others, so there is no restructuring, just different development.[1]

If one deals only with language acquisition, on the other hand, not with a historicist theory of change, and views the acquisition of new I-languages as resulting from exposure to new PLD within a speech community, then there are new structures, new parses, but no restructuring as such (Lightfoot 1999, 2006a). We now begin to consider the emergence of new structures of a wide range, sometimes yielding different parses for the same sentences, sometimes new structures that generate different sequences of words, and sometimes new meanings of words and new semantic interpretations. These are cases of sequential acquisition, where new E-language experienced by some children triggers new elements in I-languages while everything else remains constant—hence minimal differences between I-languages.

That opens the possibility of correlating new E-language phenomena with the emergence of new I-languages and thereby seeing what it takes in E-language for children to parse things differently and to acquire a new element of I-language. That is, we see what triggers the new parse. There may be changes in the ambient E-language, perhaps small changes, that lead to a new parse and trigger a new I-language that, in turn, generates significantly different expressions. The challenge for generative diachronic syntacticians is to show which elements of E-language triggered the new elements of the new I-language.[2]

We begin with a remarkable change in the history of English, when some forty psychological verbs (called "psych verbs") underwent a parallel change in meaning and a major structural shift in their morphosyntactic properties. The *parallelism* of the new phenomena points to the *singularity* of the change at the abstract level of I-languages; that, in turn, results from new parses of E-language. It is striking that English has undergone developments unparalleled in other European languages, and each of them needs to be understood in terms of children acquiring new structures that did not emerge in other language communities; this makes English a particularly interesting language to study and is presumably a function of English's external history, interacting with the languages of invading people long ago and colonized people more recently.

So-called restructuring *is* syntactic change, illustrating how I-languages sometimes vary minimally, with only one thing changing: parsing assigns new structures to E-language expressions and new I-languages develop to accommodate the new structures, yielding a multiplicity of new phenomena. We have explanations for the new parses if we can point to prior shifts in E-language that enabled the new parses. Children must have heard different things such that optimal parses differed from what happened under earlier I-languages.

3.3   Psych Verbs

Sometimes we find changes in the meaning and syntax of individual words and can explain them in terms of the abstract properties of the I-languages being acquired, that is, in terms of new parses. A meaning change, mentioned briefly in §1.3, that has intrigued historical linguists for over a hundred years concerns psych verbs such as *like*. The verb *lician* 'like' in Old English used to mean 'please', in some sense the "opposite" of its modern meaning, reversing from 'give pleasure to' to 'receive pleasure from'. So one finds expressions like (1), where 'faithlessness' is the subject of *licode* and carries nominative case.

(1) Gode ne licode na heora geleafleast.
    to.God [dative] not liked their faithlessness [nominative]
    (Ælfric, *Homilies*, vol. 2, homily 20)
    'Their faithlessness did not please God.'

Lician was a common verb: there are over four hundred citations in the *Concordance to Old English*. Denison 1990 notes that the type with a dative experiencer and a nominative theme make up "the overwhelming majority, to the extent that it is doubtful whether the others are grammatical at all."

Nor was this a phenomenon of one isolated verb. Some forty verbs occurred with a dative experiencer, usually preverbal (Lightfoot 1979: §5.1):

(2) a. *Chance*:
       At last him chaunst to meete upon the way a faithlesse Sarazin.
       (1590; Spenser, *The Fairie Queene*, book 1, canto 2, stanza 12)
       'At last he chanced to meet on the way a faithless Saracen.'
    b. *Motan* 'must':
       Vs muste make lies, for that is nede, oure-selue to saue.
       (c1440; York Mystery Plays, play 38, *The Carpenteres*, line 321)
       'We must tell lies, for that is necessary in order to save ourselves.'
    c. *Greven* 'grieve':
       Thame grevit till heir his name.
       (1375; Barbour, *The Bruce*, book 15, line 541)
       'It grieves them to hear his name.'

Jespersen 1909–1949: vol. 3, §11.2 notes many such verbs that underwent a parallel reversal in meaning: *ail*, *repent*, *become* (= 'suit'), *matter*, *belong*, and so on. So Swinburne could write *It reweth me* and *Will it not one day in heaven repent you?* but later people used personal subjects and said *I rue my ill-fortune* and *Will you not repent (of) it?* *Rue* and *repent* meant 'cause sorrow' for Swinburne and 'feel sorrow' later.

Visser 1963–1973: §34 notes that several such verbs entered the language in early Middle English: listed are *him irks*, *him drempte*, *him nedeth*, *him repenteth*, *me reccheth*, *me seemeth*, *me wondreth*, *us mervailleth*, *me availeth*, *him booteth*, *him chaunced*, *him deyned*, *him fell*, *him happened*, *me lacketh*, *us moste*, and more. This indicates that the construction was still productive in Middle English and many such verbs eventually underwent a parallel reversal of meaning.

A notable feature of psych verbs in Old and Middle English is the wide range of syntactic contexts in which they appear—sometimes impersonally with an invariant third-person singular inflection, as in (3a), sometimes with a nominative theme, as in (3b), sometimes with a nominative experiencer, as in (3c) (Anderson 1986).

(3) a. Him   ofhreow   þæs mannes.
       to.him [dative] there.was.pity because.of.the.man [genitive]
       (Ælfric, *Homilies*, vol. 1, homily 8)
       'He pitied the man.'

    b. Þa ofhreow   ðam munece þæs hreoflian mægenleast.
       then brought.pity to.the.monk the leper's   feebleness
       (Ælfric, *Homilies*, vol. 1, homily 23)
       'Then the monk pitied the leper's feebleness.'

    c. Se mæssepreost   þæs mannes   ofhreow.
       the priest [nominative] because.of.that.man [genitive] felt.pity
       (Ælfric, *Lives of Saints*, life of Oswald)
       'The priest pitied that man.'

Adopting the general framework for psych verbs of Belletti and Rizzi 1988, we may subcategorize verbs to occur with experiencer and theme DPs. Under that approach, two lexical entries typify several Old English psych verbs, linking DPs with inherent cases:

(4) a. *Hreowan* 'pity': experiencer ↔ dative; (theme ↔ genitive)
     b. *Lician* 'like/please': experiencer ↔ dative; theme

Two principles are relevant for mapping lexical representations into syntactic structures: (a) V assigns structural case only if it has an external argument (this used to be known as "Burzio's Generalization"), and (b) an experiencer must be projected to a higher position than a theme DP.

Hreowan 'pity' and several other verbs sometimes occurred just with an experiencer DP in the dative case (*Me hreoweþ*, 'I felt pity'); or they assigned two inherent cases, dative and genitive, yielding surface forms like (3a). *Lician* 'like/please' usually occurred with an experiencer in the dative and a theme in the nominative (1). With a lexical entry like (4b), the theme would receive no inherent case in initial structure. Nor could it receive a structural case from the verb, because verbs assign structural case only if they have an external argument (a nominative subject); this would be impossible, because the experiencer, having inherent case (dative), could not acquire a second case. Therefore, the theme could receive case only on being externalized, that is, on becoming a subject with nominative case.

Belletti and Rizzi 1988: §4.2 argues that dative DPs move to subject position, carrying along inherent case, therefore not receiving structural case; instead, nominative is assigned to the theme. This, in turn, suggests that dative experiencers also moved to subject position in Old English, and this seems to be true: dative experiencers showed some subject properties (this led to chaotic behavior by editors concerned to "correct" texts, discussed in Allen 1986 and Lightfoot 1999: 131).

I-languages along the lines sketched here generate a good sample of the bewildering range of contexts in which psych verbs may occur in Old English (Lightfoot 1999: 132); Anderson 1986 discusses the three examples with *hreowan* in (3) and suggests plausibly that this represents the typical situation for Old English psych verbs, that verbs attested in only one or two of the three possible syntactic contexts merely reflect accidental gaps in the texts. The writings of some writer might attest only the patterns of (3a,b) for a given verb, not (3c), but that does not entail that that writer's I-language could not generate (3c).

This is a reasonable analysis of psych verbs in Old and Middle English, and it permits a good treatment of the changes that occurred. Verbs denoting psychological states underwent striking changes in Middle English in morphology, syntax, and meaning, and we can understand aspects of these changes in terms of the loss of the morphological case system (for a fuller account, see Lightfoot 1999: §5.3). That was the change in E-language, the loss of morphological case, and we want to know what the new PLD triggered in emerging I-languages. If the morphologically oblique cases (dative, genitive) realized abstract, inherent cases assigned by verbs and other heads, their loss in Middle English meant that the inherent cases could no longer be realized in the same way. Evidence suggests that as their overt, morphological realization was lost, inherent cases were lost at the abstract level. DPs that used to have inherent case came to have structural case (nominative, accusative), and this entailed syntactic changes. After the E-language change, children ceased to acquire inherent case; inherent case played no role in their I-languages.

An I-language with oblique cases would generate forms like (3a), given the lexical entry (4a). Inherent case would be assigned to the experiencer and the theme, realized as dative and genitive respectively. Similarly, it would generate (2) and (5).

(5) a. Him   hungreð.
       him [dative] hungers
       'He is hungry.'

    b. Me   thynketh I heare.
       me [dative] thinks   I hear
       'I think I hear.'

    c. Mee likes … go see the hoped   heaven.
       me   pleases go see the hoped.for heaven
       (1557; *Tottel's Miscellany*, ed. E. Arber [London, 1870], p. 124)
       'I like to go and see the heaven hoped for.'

DPs with an inherent case could not surface with a structural case, because nouns have only one case. If a DP has inherent case, it has no reason to move to another DP position (but see Richards 2013).[3]

An I-language with no morphology realizing inherent case, on the other hand, lost those inherent cases; consequently, the lexical entries exemplified by (4) lost case specifications. Now, given a requirement that all DPs have case—the case filter of Vergnaud [1977] 2008—the DPs had to surface in positions in which they would receive structural case, governed by Infl, V, or P. This entailed that one DP would move to the subject position (specifier of IP), where it would be governed by Infl and receive structural case, nominative. A verb may assign structural case (accusative) only if it has an external argument. So the experiencer DPs in (2, 5), which formerly had inherent case realized as dative, came to occur in nominative position: *He chanced…, He hungers, I think…, I like*, and so on. With a verb assigning two thematic roles, as one argument moved to subject (nominative) position, the other could be assigned structural accusative (governed by V), generating forms like *The priest* [nominative] *pitied the man* [accusative].

Consider now the verb *like* and its lexical entry (4b) yielding experiencer [dative]–verb–theme [nominative] (1). Children lacking morphological dative case would analyze such strings as experiencer–verb–theme, with no inherent cases. Since the experiencer often had subject properties (Lightfoot 1999: 131) and was in the usual subject position to the left of the verb, the most natural parse for new, caseless children would be to treat the experiencer as the externalized subject, hence having nominative case. The new analysis of the experiencer as a subject would permit the verb to assign accusative case to the theme. And if the experiencer was now the subject, *like* would have to "reverse" its meaning, becoming 'receive pleasure from', if the sentence was to convey the same thought as forms like (1). This simply follows, given Vergnaud's restrictive UG theory of case and the postulated relationship between morphological and inherent case. Children lacking morphological case would parse (1) as experiencer–verb–theme, with *like* meaning 'enjoy', quite different from what had happened in earlier generations.

New I-languages spread through a population of speakers. There were Middle English I-languages with and without morphological case. That kind of variation entails that once some people parse what they hear with no oblique morphological case, one might find any psych verb with nominative and accusative DPs. As dative case ceases to be attested, one ceases to find verbs that lack a nominative subject—but the loss of the old patterns was gradual. The morphological case system was lost over the period from the tenth to the thirteenth centuries, so I-languages with and without inherent case coexisted for a few hundred years. The first nominative-accusative forms are found in late Old English, but impersonal verbs without nominative subjects continue to be attested until the mid-sixteenth century. The gradualness of the change is expected if the loss of inherent case is a function of the loss of morphological dative.

There is more to be said about these changes, and the reader is referred to Lightfoot 1999: §5.3. Allen 1995, Bejar 2002, and Roberts 2007: 153–161 provide good discussion relating new I-languages to the loss of morphological case but implementing the change somewhat differently. Meanwhile we can see how changes in the meaning of *like* (and other phenomena) can be explained through the loss of inherent case, an automatic consequence of the loss of oblique morphological cases. As morphological cases disappeared, an E-language change due to Scandinavian influences (see Van Kemenade 1987 and O'Neil 1978), the new I-language without inherent case was the only option if everything else stayed the same.[4] In this way we gain an impressive depth of explanation, showing that new E-language with no morphological case was parsed differently, with no inherent case and with quite different structures for expressions with experiencer subjects. New E-language required different parses, therefore different I-languages that generated quite different structures and new meanings for a significant class of verbs.

We can understand why and how these changes took place, how children invented new I-languages on the assumption that they were parsing new E-language. Meanwhile it is hard to see how this diversity of phenomenological change could be stated in terms of a new setting of a binary parameter defined at the level of UG, given that it involves not only new structures but also new morphology, as well as new meanings for certain specific verbs.

3.4 Chinese *Ba* Constructions

Typologists of the 1970s argued that Chinese languages had undergone major structural shifts; for example, Li and Thompson 1974 argues that basic word order changed from object–verb to verb–object and back again to object–verb. Li and Thompson's arguments are based on harmonic principles based, in turn, on the word-order generalizations of Greenberg 1966, but their analyses involve many complexities that would raise fundamental problems for any translation into a generative treatment in terms of parameters, given that complex parameters would need to be postulated as part of UG.[5] Whitman 2001 and Whitman and Paul 2005 express skepticism. A particular bugbear of Chinese specialists has been what are known as *ba* constructions. For a textbook treatment, see Huang, Li, and Li 2009: §5, which addresses the difficulties in determining the best structures.

In early Chinese, *ba* first occurred as a serial verb. Whitman and Paul adapt Larson 1991's and Collins 1997's VP-complementation analysis of serial verbs and assign structures like those of (6) (further adapted here to incorporate a Copy-and-Delete analysis of what used to be treated as movement).

(6) a. Sunzi $_{v'}$[jiang/ba $_{VP^1}$[yi-ya $_{V''}$[~~jiang/ba~~ $_{VP^2}$[si $_{VP^2}$[*pro* yong]]]]].
    Sunzi take   one-duck   privately use
    'Sunzi grabbed a duck and used it himself.'

b.

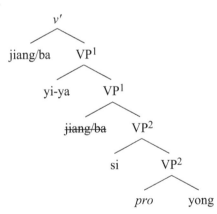

In (6) *ba* (with its lexical meaning 'take') has as its complement the second VP headed by the verb *yong* 'use'. The shared object *ya*

'duck' is merged in the specifier of the VP headed by *ba* and controls *pro* in the complement VP. *Ba* is copied to *v*, deriving the surface order. The first, higher verb takes as its complement the VP headed by the second verb; the "shared" object of the serial construction originates in the specifier of the first, higher verb and the surface V^1–object–V^2 order is derived by copying V^1 around the object to *v*. Under this analysis, the first uses of *ba* were serial-verb constructions of the kind found in many African languages; indeed there have been productive collaborations between Sinologists analyzing *ba* and other such elements and Africanists analyzing serial verbs. Whitman and Paul (n. 9) point to Carstens 2002's analysis of serial verbs in Yoruba and other Niger-Congo languages as a model for their analysis of early *ba* constructions. If there were a parameter for serial verbs, one might say that it was set positively in Middle Chinese, but we shall see that there is little prospect of a unified treatment of serial verbs manifesting a single structural property of the kind that a parameter might embody.

So historically, *ba* was a lexical verb meaning 'take', 'hold', or 'handle' and appeared in "serial-verb constructions," V^1 [*ba*+NP+V+XP], which might mean 'take NP and do [V+XP] to it', like the examples in (7). We will see that this analysis does not hold for modern Chinese.

(7) a. Wo ba juzi   bo-le   pi   le.
        I   BA orange peel-LE skin LE
        'I peeled the skin off the orange.'
    b. Ta   ba Lisi paoqi-le.
       3SG BA Lisi abandon-PERF
       'She abandoned Lisi.'

Huang, Li, and Li 2009: 153 characterizes some basic facts about *ba* in *modern* Mandarin: "The object of *ba* is typically … the object of a verb," and "this object is 'disposed' or 'affected' in the event described": for example, 'that scoundrel' in (8a). Likewise the noun *ma* 'horse' after *ba* in (8c) must be made tired by the riding, excluding the second meaning of (8b).

(8) a. Lisi ba na-ge   huaidan   sha-le.
       Lisi BA that-CL scoundrel kill-LE
       'Lisi killed that scoundrel.'
    b. Linyi qi-lei-le   ma.
       Linyi ride-tired-LE horse
       i. 'Linyi rode a horse and made it tired.'
       ii. 'Linyi became tired from riding a horse.'
    c. Linyi ba ma   qi-lei-le.
       Linyi BA horse ride-tired-LE
       Intended meaning: same as (8bi)

There is a vast literature on the complexities of this construction, for example, Chao 1968; Peyraube 1985; Sybesma 1999. Beyond straightforward cases like these, accepted by everybody, there is a wide range of extended examples whose acceptability varies across different dialects. The semantic restriction seen in (8) has been attributed to a general requirement that the post-*ba* NP be affected by the action of the verb.

However, in modern Mandarin, from Middle Chinese onwards, *ba* has lost the usual lexical-verb properties, having become "grammaticalized," rather as the modal verbs were grammaticalized to Infl elements in Early Modern English, as discussed in §2.4. Like those modal verbs, *ba* differs from lexical verbs in that it cannot take an aspect marker as in (9a), cannot form an alternative V-not-V question like (9b), cannot assign a theta role (it lacks lexical feature content, e.g., the feature associated with a theta role), and cannot serve as a simple answer to a yes–no question like in (9c).

(9) a. *Ta ba-le   ni   hai(-le).
       he BA-LE   you hurt-LE
       'He hurt you.'

    b. *?Ta ba-mei/bu-ba ni hai(-le).
       he BA-not/not-BA you hurt-LE
       'Did he hurt you?'
    c. *(Mei/bu-)ba.
       not-BA

Saying that *ba* has been grammaticalized raises questions about what exactly that means: what are the morphosyntactic properties of a grammaticalized *ba*? Many possibilities have been put forward.

Huang, Li, and Li 2009: 174 summarizes the major properties of *ba* constructions as in (10); the central property is that *ba* is no longer a lexical verb.

(10) a. A *ba* sentence is possible only when there is an inner object or an outer object. The post-*ba* NP is an inner or outer object—but not the outer*most* object.

     b. Although *ba* assigns Case to the post-*ba* NP and no element can intervene between them, they only form a syntactic unit in "canonical" *ba* sentences and not in causative *ba* sentences.

     c. *Ba* does not assign any theta roles: neither the subject of the sentence nor the post-*ba* NP receives a theta role from *ba*.

     d. The *ba* construction does not involve operator movement.

The generalizations of (10) need to be captured by the structures postulated, that is, by the parse that children make; that is the fundamental challenge. Whitman and Paul 2005: (16) hypothesizes the following structure.[6]

(11)

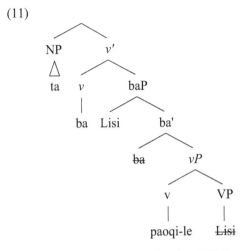

she   Lisa   abandon-PERF

'She abandoned Lisi'

What changes in modern *ba* constructions is that *ba*, which orig-
inated as the serial verb 'take', comes to be base generated in the
head of the higher functional category (Whitman & Paul do not
specify what that category is), and, related to that, *ba* ceases to
assign a theta role. As a result, after the new parse, the object must
be introduced as the complement of V^2 and copied to the specifier
of the projection of *ba*. As earlier, the surface word order must be
derived by copying *ba* to a higher functional-head position, namely
v in (11).

If new parses emerged in Middle Chinese along these lines, one
wants to know why, what new E-language phenomena required the
new parses. This requires the competence of a specialist in Chinese
syntax, and John Whitman has generously shared some interesting
suggestions (personal communication).

One factor is that earlier Chinese had another "disposal" construction of the same type, *jiang*, also meaning 'take, hold', illustrated in (12) (Whitman & Paul 2005: 3). Whitman notes that *jiang* still survives in the same function in southern varieties of Chinese but that *ba* has replaced *jiang* in northern varieties like Mandarin.

(12) Sunzi jiang yi-ya   si   yong.
      Sunzi take   one-duck privately use
      (Eighth century; Zhang Zhuo, *Chao ye quian zai*)
      'Sunzi grabbed a duck and used it himself.'

Second, after Old Chinese (roughly third–fourth century CE), later Chinese began to disallow definite bare NPs to the right of lexical verbs. Old Chinese is more consistently verb–object than later varieties. *Ba/jiang*-marked complements are the device for realizing definite/specific bare-NP complements. This is a functionalist account, but that is entirely appropriate as a description of E-language phenomena. The new development may be conditioned by contact with "Altaic" languages such as Turkic, Mongolian, and Tungusic. All of these are object–verb languages with some degree of differential object marking: as in modern Turkish, bare objects on the right are nonspecific, while objects on the left with overt accusative marking are specific/definite. There was massive bilingualism, especially in the north, between Chinese and all three of these language groups starting prior to the Tang period, essentially the advent of Middle Chinese.

A third factor affecting available parses is a general decline in lexical serial-verb constructions; such forms would have required the earlier parse (6), construing *ba* as a lexical verb. They survive, Whitman notes, in double-object constructions but rarely elsewhere.

It is clear that in modern Mandarin *ba* is a functional head. Generally, *ba* and the post-*ba* NP do not form a constituent; hence the structures posited in (11) and note 6. *Ba* does not assign a particular theta role, so the post-*ba* NP can be any NP, copied from anywhere, to a position immediately to the left of the finite verb in second position. *Ba* does assign case, however, and is necessarily followed by an NP/DP; consequently *ba* must be a light verb. The post-*ba* NP is generally the affected NP/DP, the outer object assigned a theta role by a complex predicate, V', consisting of a verb and its complement (in note 6). That structure also captures the fact that *ba* follows the aspectual markers *you* and *zai*, which head Aspect Phrases below IP but above VP, as Huang, Li, and Li 2009: §3.3.1 shows.

It is hard to imagine how any UG-defined, binary parameter might illuminate the analysis of modern *ba* constructions and explain their emergence. Specifically, if the early *ba* constructions manifested a lexical, African-style serial-verb parameter, one wonders how the new, modern constructions would be treated by a parameter analysis, when the indications are that *ba* forms nowadays manifest a functional category. Merely saying that *ba* is "grammaticalized" would not explain the idiosyncrasies associated with such forms—more specific information is needed.

On the other hand, a parsing analysis would have children discovering that *ba* is base generated as a functional category assigning case but no theta role to an affected NP, and that seems much less problematic, appearing to be straightforwardly learnable on the basis of data available to young children. Data triggering these structures are readily available to children parsing their ambient language. It is difficult to pinpoint exactly what shifts in E-language triggered the new parse, but Whitman's suggestions are of the right kind; they show, for example, how E-language shifts reduced the expression of the old parse (6, 7), treating *ba* as a lexical (serial) verb. The story, however, remains incomplete until we learn how E-language changed in such a way that the trigger for the old analysis no longer functioned in the same way and how new E-language triggered the new parses. Chinese-speaking readers might like to investigate changes taking place in E-language just prior to the emergence of the new analysis and explore hypotheses about the before and after parses.

Meanwhile we have learned that *ba* changed from a lexical, serial verb to a functional category, a light verb. It is impossible to see how this could be treated as a change in a UG-defined parameter setting. UG-defined parameters were supposed to harmonize E-language phenomena, showing how clusters of phenomena were unified under a single parameter setting. But variation seems not to work in the expected way. The grammaticalization of English modal verbs discussed earlier resembles the grammaticalization of Chinese *ba* in some ways but not in all ways, so merely pointing to grammaticalization does not suffice. The associated phenomena differ: Chinese *ba* constructions single out NPs that are affected by the action of the verb phrase, but that plays no role in the behavior of English Infl elements. In addition, complements of English Infl elements are distinctive in ways that complements of Chinese *ba* constructions are not.

Successful parsing depends on discovering the right contrasts, as explored in much work emanating from the Toronto school—Dresher, Cowper, Hall, and others (see, e.g., Dresher 2009, 2019; Cowper & Hall 2019). At certain points English lexical verbs became distinct from their modal counterparts, and *ba, jiang,* and so on became distinct from other lexical verbs and came to pattern like light verbs. The new parses adopted by children involved other complexities that differed in each of the languages. It looks as if UG-defined parameters are too gross to capture the variable properties found in natural languages. It appears that the variable properties associated with Chinese *ba* constructions are best understood as resulting from the parsing of new E-language on the part of young children—different E-language from that triggering the parses of Infl elements in English.

Needless to say, in this section we have just scratched the surface of the behavior of Chinese light verbs, which is more complex than we have seen; that simply makes a unified parameter analysis still more problematic.

3.5   Null Subjects

What emerges from our consideration of Chinese light verbs is that it is difficult to see a unified analysis of the disparate phenomena, as would be required for an analysis in terms of a single parameter defined at UG. Instead, children seem to discover various structures piecemeal through parsing the ambient language. Similarly with null subjects.

It is worth noting that there tends to be a striking asymmetry between discussion of UG principles and UG-defined parameters: the former are almost always based directly or indirectly on poverty-of-stimulus arguments and are shown to account for negative data, information about what does *not* occur. For example, the binding principles are motivated by the nonoccurrence of structures like *Sally$_i$'s father washed herself$_i$*, *Sam$_i$ hurt him$_i$*, and others; conditions on deletion block the generation of *Who do you think that ~~who~~ said this?*; structure-dependence conditions disqualify *Is the ball that ~~is~~ red is hard?*, as distinct from *Is the ball that is red ~~is~~ hard?*

On the other hand, in discussion of parameters, the role of comparable negative data is more ambiguous. For example, if there is an object–verb-order parameter, depending on how it is formulated, it is not necessarily clear that this parameter is needed to block anything that doesn't occur. If there is a null-subject parameter, perhaps it bars the generation of null subjects if verbal morphology is not rich enough, in English or German for instance, but such generalizations are undermined by apparent null subjects in Chinese, which require a different analysis.

In that context, it may be helpful to reconsider the parade case of a UG-defined parameter, the null-subject or "*pro*-drop" parameter, characterized in Chomsky 1981a; Rizzi 1982: c143 in terms of referential null subjects that co-occur with free subject–verb inversion in simple clauses; rich verbal morphology; null resumptive pronouns in embedded clauses; and apparent violations of the *that–trace* filter.

An interesting context is Portuguese: European Portuguese is usually analyzed as a fairly standard Romance null-subject language like Italian or Spanish, while Brazilian Portuguese is regarded as a partial null-subject language from the nineteenth century onwards, when overt subjects became increasingly common. However, new work by Humberto Borges and Acrisio Pires (2017), examining 2,500 clauses from the oldest available corpus of diaries/journals written in Goiás in the central-west region of Brazil, shows that loss of null subjects began earlier there. Realization of overt subjects began to be more common in the eighteenth century. Also, verb–subject inversion, typically taken as a hallmark of null-subject languages, drops from 57 percent in the eighteenth century to only 22.5 percent in the nineteenth century. However, verbal morphology did not become impoverished as null subjects became less common, as one would expect if there were a unified null-subject parameter keyed to rich verbal morphology, as is often argued (e.g., Rizzi 1982: chap. 4). This suggests that, as with light verbs in Chinese, we may be cutting the pie into pieces that are not small enough, seeking to unify phenomena that are better treated piecemeal and independently. Specifically, it looks as if the new parse, whatever it was, supplanted the analysis allowing subject–verb inversion. Whatever the correlations, one wants to know what the new parse consists in and what new E-language phenomena triggered it. This is a matter that Borges and Pires are turning to in their current and future work. Thinking in terms of new parses triggered by new E-language phenomena changes what one looks for in the historical record.

Similarly, Cecilia Poletto (2018) reports studying the distribution of null subjects in thirteenth-century Old Italian, seeking to discover the factors influencing their distribution in that language. She shows that a discrepancy noted long ago for Old French (Adams 1987), that they are frequent in main clauses and rare in embedded domains, holds in Old Venetian (except in interrogatives) but not in Old Florentine, sometimes referred to as Old Italian, where there is no such discrepancy between main and embedded clauses. This means that the direct link that has sometimes been assumed between V-to-C movement (subject–verb inversion) and null subjects cannot be maintained.

Poletto shows that there is no clear-cut asymmetry between main and embedded clauses in Old Italian as there is in Old French and Old English, since null subjects are also numerous in embedded domains. In addition, the syntax of null subjects is not the same in Old Italian and modern Italian (despite the protestations of Zimmerman 2012), since in Old Italian embedded clauses subject pronouns occur where there are obligatory null subjects in modern Italian. Poletto also rejects the possibility that null subjects in embedded domains depend on the fact that Old Italian was a symmetrical verb-second language, since Old Italian embedded clauses do not show verb-second characteristics.

In short, conventional analyses of the null-subject parameter unify things that behave quite differently, and therefore they seem not to be cutting the pie into good slices; this, then, is another area where rethinking is needed. For excellent discussion of the null-subject parameter, see Duguine, Irurtzun, and Boeckx 2017. The authors, like me, view UG-defined parameters as incompatible with the aspirations of the Minimalist Program and provide empirical evidence that parade instances of parameters, the null-subject parameter and the compounding parameter, are problematic, showing that other ways of capturing variable properties are very much needed. In particular, they argue that these parameters do not solve Plato's problem, the logical problem of language acquisition, as I am also arguing here. Furthermore, they show that these parameters fail to capture the right clustering of properties: empty nonreferential subjects, free subject–verb inversion, absence of *that*–trace effects (Rizzi 1982). They note that "the typological correlations between pro-drop and these other grammatical properties have been shown to be quite weak descriptively" (p. 447) and that the null-subject parameter fails to cut the pie in productive or insightful ways.

Duguine, Irurtzun, and Boeckx's opening paragraph lays out the general questions clearly: Are crosslinguistic variable properties constrained, and is UG rich enough to contain principles that contain variables that must be set in the course of development? Positing parameters answers those two questions in the affirmative. But if there were a (macro)parameter for "*pro*-drop" or "null-subject" languages, some property of E-language would determine whether a language is *pro*-drop or not, and this would be a monolithic, defining property of both language types. Duguine, Irurtzun, and Boeckx argue that this is not the case. If they are right, the correct analyses fly in the face of the fundamental idea of the parameter vision (§1.2), that variation between languages consists of universal clusters of phenomena that show up repeatedly in crosslinguistic work.

Recent comparative work militates against that vision (for good general discussion, see Cinque 2013). First, non-*pro*-drop languages sometimes allow null subjects in certain contexts, for instance, in Diary British English (Haegeman & Ihsane 2001). And the inverse relation also holds. For instance, Basque, a null-subject language, does not allow null subjects in nonfinite adjuncts:

(13) [Zu/*e hainbeste mintza-tze-a-n]   (ni) nekatzen naiz.
     you   so.much speak-NMLZ-DET-LOC] I   tire   AUX
     Lit. '[When you/e talking so much] I get drained.'

The granularity of null subjects is also a problem. Many languages show mixed properties, cases where null subjects are possible and cases where they are not possible. Partial null-subject languages would have some sort of in-between status with respect to the (macro)parameter.

In addition, there are many ways in which the granularity problem arises, since different languages mix *pro*-drop and non-*pro*-drop in different dimensions. Some languages display a split with respect to person features (Bavarian or Marathi), but the split can also be related to tense (Hebrew). Going beyond subjects, we see languages with null objects as well. Sometimes a language has both null subjects and null objects (Basque, Japanese), others have null subjects and not null objects (Spanish, Italian), and others are the other way round (Brazilian Portuguese). Duguine, Irurtzun, and Boeckx conclude: "In sum, not only is pro-drop not binary, it is not even discrete"; the availability of null subjects is construction-specific, not specific to a whole language, and the distribution of null subjects is not what one would expect under a macroparametric analysis (p. 451).

3.6   Reflections

In this chapter, I have joined with Bob Berwick (1985) and Janet Fodor (1998c) in viewing parsing as a key ingredient in language acquisition and in the analysis and explanation of variable properties. A child identifies the words, categories, and structures of their I-language through parsing, they accumulate the structures by which they come to be able to produce and understand an infinite range of thoughts. I go beyond Berwick and Fodor: given a highly restrictive Minimalist theory of UG and a view of parsing as implemented by the emerging I-language, children are driven to a narrow range of possible structures that they select from in parsing. Since Merge is an important element in children's toolboxes, structures must be hierarchically organized and binary branching. A head may combine with a complement, specifier, or adjunct. Similarly, heads may project into phrasal categories containing head–complement, adjunct–head, and so on. That much follows from possibilities defined by UG. The possibilities for parsing are limited given that parsing is implemented by I-languages.

Likewise children may understand *What garage did Kim park the car in?* to mean 'Kim parked the car in a garage; which one was it?' but not **What garage did Kim like the car in?* to mean 'Kim liked the car in a garage; which one was it?', because the latter violates UG locality restrictions. *Park* allows a structure $_{V'}[_{V}park$ $_{DP}the\ car]$ $_{PP}[in\ a\ garage]$, but *like* allows only a complex DP as its complement: $_{DP}[the\ car\ in\ a\ garage]$, not $_{DP}[the\ car]$ and $_{PP}[in\ a\ garage]$. The locality constraints prevent what *garage* (replacing *a garage*) from being copied from inside this complex DP to the edge of the sentence. Put differently, children learn that *park the car in a garage* has a structure $_{VP}[_{V'}[_{V}park\ _{DP}the\ car]\ _{PP}[in\ a\ garage]]$, and this is reflected in the fact that they readily accept and produce *What garage did Kim park the car in?*, since it violates no locality constraints.

Other things follow from the emerging I-language. For example, at a certain point a Dutch-speaking child might have an I-language where prepositions precede their complement (*over de water* 'over the water') and then develop a further restriction that a preposition follows its complement if it is an *-r* pronoun (*daar over* 'thereover') (Van Riemsdijk 1978).

We have shown how children might acquire a structure accommodating expressions like this; such structures are triggered by their ambient E-language. Triggering such structures seems to offer a more tractable kind of learning than setting parameters by evaluating the generative capacity of hypothesized grammars. Indeed, given the restrictions of the invariant principles, it is imaginable that parsing is done by children equipped biologically with those invariant principles and an emerging I-language. Under this view, there would be no parser independent of the invariant principles of UG and of partially formed I-languages. The parsing capacity is subject to the conditions of UG, the partially formed I-language, and the interface conditions—the subject of the next chapter.

If children acquire their eventual I-language by parsing their ambient E-language and postulating the I-language structures needed to do so, then linguists can dispense with parameters defined at UG, allowing only invariant principles at that level, a major simplification compared to analyses consisting of both principles and quite distinct parameters (Chomsky 1981b).

Shedding an independent parser and the inventory of parameters defined at UG represents an intriguing possibility. We have shown what kind of analyses that possibility might lead to; we will need more good stories of E-language shifting in such a way as to trigger new I-language structures, and some current stories will need to be fleshed out beyond what we have so far, for example, concerning Chinese *ba* constructions (§3.4) and the new analyses of forms of *be* that occurred in nineteenth-century English (§2.6).

Meanwhile interesting work is proceeding on ideas about an independent parsing capacity. For example, Luigi Rizzi and Guglielmo Cinque (2016), developing their rich "cartographic" theory of functional categories, have pointed to the usefulness of such ideas for psycholinguists working on the acquisition of lexical items and the fixing of syntactic parameters. Similarly, Heidi Getz (2018b) has done impressive work on how children learn apparently complex things (Getz 2018a, 2019), and she has revived ideas about the special role of closed-class items in parsing and in acquisition (e.g., Emonds 1985; Valian & Coulson 1988 on "anchoring"). If closed-class items have a functional role to play in identifying basic syntactic categories, one wonders why they appear relatively late in children's speech and why there are languages that seem not to privilege these items. At this time, the role of closed-class items remains a mystery and we need to stay open to a range of possibilities.

Under the view taken here, variation does not follow from the narrow prescriptions of a set of parameters defined at UG but rather is contingent; dependent on what history has wrought, how features of E-language have triggered the selection of elements of I-language in the private systems of some members of a speech community; and free of any independent, formal conditions. This allows us to define language variation more broadly and to offer quite different, more contingent explanations for what comparativists have observed. UG is open in at least two distinct senses: first, it allows words to be categorized differently from language to language (*must* is an Infl in modern English, but its translation, *devoir*, is a verb in French I-languages); second, parsing fleshes out a child's I-language, in great detail, enriching the structures that a child selects, as we have seen at several points in this book.

We will return to these considerations in chapter 5, after considering the role of interfaces.

4 Parsing at Interfaces

4.1 Logical Form: Pronouns

So far we have been discussing syntactic well-formedness in terms of what UG prescribes about syntactic structures and what is learned by children through the parsing process, that is, through discovery and selection. However, in the most general terms, languages are systems in which syntax connects meaning to some form of EXTERNALIZATION. Meanings, as framed by logicians, philosophers, and linguists, are "logical forms," consisting of units formed from subunits, head–complement relations, phrase–adjunct relations, thematic roles, indexical relations, topic–comment relations, presupposition–assertion distinctions, and much more. For every syntactic structure, there is a logical form, specifying aspects of the meaning, associated with an externalization.

In most people, the syntax connects meaning and sound: the externalization consists of a phonological form for the expression. Thus, in these individuals, for every syntactic structure there is a corresponding externalization that specifies aspects of the sound and a logical form that specifies aspects of the meaning. These are interpreted at the "sensorimotor" and "conceptual-intentional" interfaces respectively, which have their own well-formedness conditions. Those well-formedness conditions interact with learned, variable properties and therefore involve parsing on our approach. For a significant minority, the externalization is some kind of signed system, not based on sounds, but on gestures like those of American Sign Language or the new Nicaraguan Sign Language. For smaller minorities, the externalization might be tactile; the Tadoma method is one such system. Whatever the externalization used, sound, sign, or touch, the point is the same all around. Phonologists, beginning in the early twentieth century, did fundamental work developing rich descriptive systems based on the theories of contrast and distinctive features of Roman Jakobson, Edward Sapir, Nikolai Trubetzkoy, and others (Jakobson 1941; Sapir 1925; Stokoe 1960; Trubetzkoy 1939). In the 1960s, analysts followed the seminal work of William Stokoe of Gallaudet University in Washington, DC, and set about understanding how signed languages conveyed the meanings that syntacticians had discovered in working on oral languages. Soon a vibrant research community emerged, finding that signed systems were as rich and complex as oral systems and making fundamental discoveries about individual signed languages, which proved to bear essentially no relation to their ambient oral languages.

In this chapter we will consider some difficult ellipsis phenomena that interact with both the sensorimotor interface and the conceptual-intentional interface and have interesting logical and phonological consequences. They also do not manifest the properties one would expect if linguistic variation were due to binary parameters defined in UG. Let us begin by considering the conceptual-intentional or syntax–meaning interface, in particular the Binding Theory and the parsing issues that it raises. The logical form will involve the *interpretation* of pronouns, particularly those in ellipsed VPs, such as *Papa Bear wiped his face and Brother Bear did* ~~wipe his face~~, *too.* Then in the next section we will consider the *distribution* of ellipsed VPs, determining where VPs may be reduced to silence in this way.

The early history of generative grammar spawned many publications on the way in which pronouns could refer to people and objects. Linguists offered complex indexing procedures, whereby DPs might have the same or different indices, depending on whether they referred to the same or different entities. For a taste of the kind of unpleasant complexity invoked, see the appendix to Chomsky 1980 and the indexing procedures postulated there. Fortunately, the classical Binding Theory was introduced soon after, in Chomsky 1981a, and consisted of just three principles, simplifying matters enormously:

(1) A. Anaphors are bound locally.
    B. Pronouns are free locally.
    C. Names are free.

To be bound locally meant being coindexed with a higher, C-COMMANDING expression that is local, within its DOMAIN; being free meant not being coindexed with a local c-commanding expression. Those three principles, simply known as Principle A, Principle B, and Principle C, permitted a dramatic simplification of analyses. By hypothesis, they constitute a component of UG, available to humans in advance of experience, in fact enabling children to interpret their experience. The principles facilitate structures that can meet the demands of learnability, because children only need to learn which nominals are anaphors and pronouns, which does not look difficult. The principles divide nominals into three types: anaphors like the reflexive pronouns *himself, themselves*, and so on (in English), pronouns like *she, her, their*, and names (all other nominals). Each nominal is contained inside a Domain, roughly its clause or a larger DP, and it is either coindexed with another, higher DP within that Domain or not. If so, then it is bound locally; if not, then it is free locally. The Binding Theory, in modern terms, involves the indexical relations that make up a well-formed logical form; it is part of the conceptual-intentional interface.

As one way of visualizing things, for an expression *Kim's mother washed herself*, whose structure is represented in (2), an analyst might start from *herself* and proceed up the hierarchical structure from one node to another. If a node is reached on this upward trajectory whose sister is a DP, that DP is a potential binder for *herself*. This approach involving tree climbing and sister checking enables us to capture the basic technical notion of c-command. The analyst gets as far as the lower IP, at which point there is a DP that is a sister to this node: *Kim's mother*.

(2)

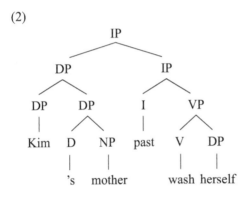

Herself must be coindexed with (and refer to) this maximal DP *Kim's mother*; it may not refer to the lower DP *Kim*, because that lower DP is not a sister to the IP—it is contained within the larger DP and is therefore inaccessible to the Binding Theory.

The representation in (3a) is an alternative and partial representation of (2).

(3) a. DP[DP[Kim]'s mother]$_i$ washed herself$_i$.
     b. DP[DP[Kim]'s mother]$_j$ washed her$_i$.
     c. DP[DP[Kim]'s mother] said CP[that the doctor$_i$ washed her$_j$].
     d. DP[DP[Kim$_i$]'s mother]$_j$ said CP[that the doctor$_i$ washed Kim$_i$].

e. Kim said $_{CP}$[that the doctor$_i$ washed her$_j$].

f. Kim$_i$ said $_{CP}$[that the doctor washed Kim$_j$].

Now consider (3b): it has the same structure, just with *her* in place of *herself*—and *her* may not be coindexed with the DP *Kim's mother*, because as a pronoun it needs to be free in its clause. It may, on the other hand, be coindexed with *Kim*, precisely because the DP *Kim* is not a sister to the IP (or any node reached by moving up the tree node by node starting with *her*) and is, therefore, irrelevant to the demands of the Binding Theory. Next, note that (3c) is ambiguous: *her* may refer to *Kim* or to *Kim's mother*. The Binding Theory stipulates only that *her*, a pronoun, be free within its own Domain, the clause (CP) indicated; beyond that, there is nothing systematic to be said and any indexing is possible. Similarly, in (3e) *her* may be coindexed with *Kim*, because *her* is thus free (not coindexed with anything) within its own clause; or it may have its own unique index, referring to a woman other than Kim. Turning to names, note that the difference between them and pronouns is that while pronouns only need to be free locally, the need of names to be free is not limited to their own Domain. On the one hand, (3d), with its two *Kim*s, can be a statement about one Kim; the lower *Kim*, the complement of *washed*, may not be coindexed with any sister DP we meet as we work our way up the tree structure (not stopping at the CP node but continuing to climb), but the higher DP *Kim* is not a sister to any node dominating the lower *Kim*, hence invisible to the Binding Theory. On the other hand, (3f) necessarily concerns two Kims; the lower *Kim* may not be coindexed with the higher *Kim*, whose DP is a sister to the IP node dominating the lower *Kim*.

The Binding Theory yields the necessary distinctions beautifully but itself cannot be learned from data accessible to young children, the PLD. We therefore say that it is part of UG, part of what children bring to the analysis of initial experience. Learning is involved, however: children must learn which words are anaphors and which are pronouns, but nothing more complex is needed. The three possibilities are defined in (1) and they hold for all languages. Once a child acquiring English has learned that *themselves* is an anaphor, *her* a pronoun, and so on, all the appropriate indexing relations follow, with no further learning. Similarly for other languages, children learn which words are anaphors and pronouns and everything else follows. How, then, do they learn which words are which? We will see that parsing must be involved.

Exposure to a simple sentence like (4a), interpreted with *themselves* referring to (coindexed with) *they*, suffices to show that *themselves* is an anaphor and not a pronoun or a name; pronouns and names may not be thus coindexed with an accessible phrasal category within their Domain.

(4) a. They$_i$ washed themselves$_i$.
    b. Kim$_j$'s father loves her$_j$.
    c. Kim$_i$ heard $_{DP}$[Bill's speeches about her$_i$].
    d. Kim left.

The sentence in (4b), interpreted with *her* referring to *Kim*, shows that *her* is no anaphor, since it is not coindexed with any sister DP encountered as we move up the tree structure within *her*'s Domain. And (4c), with *her* referring to *Kim*, shows that *her* is not a name, since names may not be coindexed with a sister DP anywhere; the Domain of *her* is the DP indicated and *her* is free within that Domain, happily. If neither an anaphor nor a name, then *her* is a pronoun. So far, so good but here comes the snag.

A very simple expression like (4d) shows that *Kim* is not an anaphor, but there is no positive evidence available to a child showing that *Kim* is not a pronoun. Analysts know that *Kim* is not a pronoun, because one does not find sentences like *Kim said that Kim left*, with the two *Kim*s referring to the same person, but that is a negative datum, information that something doesn't occur, hence unavailable to young children. So a complication has arisen; but it can be resolved by appealing to hierarchical organization.

If we turn to hierarchical relations, the starting point for a child might be that every word is a name, unless there is positive, refuting evidence. Under that view, sentences like (4a) show that *themselves* is not a name, and not a pronoun either, hence an anaphor. And (4c) shows that *her* is not a name, because it is coindexed with an accessible sister DP (*Kim*), and not an anaphor, because it is not locally coindexed, hence a pronoun. This yields a satisfactory account. We have a theory of mature capacity that provides the appropriate distinctions, and one can show how children learn from environmental data which elements are anaphors (1A) and which are pronouns (1B); all other nominals are names (1C).

However, further work suggests that this problem with the learnability of pronouns may be symptomatic of a bigger issue. Principles A and C, applying to anaphors and names, have stood the test of time well, but Principle B has been problematic from the early days of the Binding Theory. For example, Avrutin and Wexler 1992 observes that Principle B seems to be delayed in Russian-speaking children and does not come into effect until well after Principles A and C. Avrutin and Wexler argue that the effect is illusionary and that, in fact, Russian children behave according to Principle B from the earliest stage but lack a particular pragmatic principle, which they spell out (below). That is what makes it appear that Russian children lack Principle B.

According to Grodzinsky and Reinhart 1993, Principle B applies only to pronouns that are bound variables, hence to the pronoun in *Is every bear touching her?* but not to the pronoun in *Is Mama Bear touching her?*, which can be referential. A clear instance of a non-referential pronoun is *No bear likes his father*, where there can be no referent for *no bear* and therefore none for *his*. If Principle B does not apply to referential pronouns, one may wonder why *Mama*

Bear is touching her does not mean the same as the corresponding sentence with an anaphor, *Mama Bear is touching herself*. For that reason, Grodzinsky and Reinhart invoke a special rule, their Rule I: "NP A cannot co-refer with NP B if replacing A with C, C a variable A-bound by B, yields an indistinguishable interpretation" (p. 70). The status of such a rule is quite unclear; for good discussion, see Elbourne 2005.

And, talking of Elbourne, he claims (p. 361) that languages vary in terms of whether they observe Principle B: earlier forms of English and Maori, for example, do not. That leads him to conclude that Principle B is subject to a parameter and that children have to learn it only if it applies to their I-language. However, as Elbourne notes, there is a serious problem with that proposal: as we noted four paragraphs back, learning the principle would require access to negative evidence, information that certain things do not occur in the language.

Also contributing to this literature on the special properties of pronouns, Thornton and Wexler 1999: chap. 2 surveys empirical investigations showing that children have Principles A and C but not Principle B. Thornton and Wexler (p. 9) find that Principle B "stands out as an empirical problem area," as argued by Elbourne. They adapt the proposal of Avrutin's dissertation (Avrutin 1994) and distinguish three analyses of *Mama Bear is washing her face*, what they call the deictic, coreference, and quantificational readings. In the deictic reading, Mama Bear washes somebody else's face, perhaps Snow White's, and there is no coindexing. In the second reading there is coreference between Mama Bear and the pronoun. And the third reading involves a quantificational analysis with a LAMBDA OPERATOR: Mama Bear (λx (x is washing x's face)) (cf. the bound-variable use of pronouns discussed above). The second and third analyses have the same truth conditions and are difficult to tell apart. There has been a vast amount of work on the interpretation of pronouns and the circumstances under which they may be interpreted as coindexed with a preceding noun phrase; Elbourne 2005 has wise discussion.

Thornton and Wexler argue that children's misinterpretations of sentences like *Mama Bear is washing her*, with *her* referring to Mama Bear, are not violations of Principle B but reflect incomplete pragmatic knowledge. "As a consequence, children accept coreference between a pronoun and a name, what we will be calling a *local coreference interpretation*, in circumstances in which an adult would not" (p. 14). They claim that children have difficulty evaluating other speakers' intentions, which has consequences for both speech and understanding of language. Children typically take new information to be old information, in nonadult fashion, explaining why "children may announce 'He hit me' instead of 'A boy hit me' or 'John hit me'" (p. 15). This is why children allow local-coreference interpretations for expressions like *Mama Bear is washing her*.

Thornton and Wexler experiment with VP-ellipsis constructions. They compare children's interpretation of pronouns in simple sentences like (5a), governed by Principle B, with their responses to sentences like (5b), governed by Principle C. As noted, children sometimes let the pronoun and the name co-refer in (5a), as adults would never do.

(5) a. Mama Bear is washing her.
    b. She is washing Mama Bear.

Thornton and Wexler also compare children's interpretation of (5a) with their interpretation of pronouns in ellipsed VPs. An example of VP ellipsis is given in (6a); in (6b), the ellipsed VP contains a pronoun.

(6) a. Papa Bear ate pizza and Brother Bear did ~~eat pizza~~, too.
    b. Papa Bear wiped his face and Brother Bear did ~~wipe his face~~, too.

The pronoun in (6b) is multiply ambiguous and may have a deictic, coreference, or bound-variable reading. In simple sentences the coreference and bound-variable readings are hard to distinguish, because they are true under the same truth conditions, as we noted in our discussion above of *Mama Bear is washing her*. In ellipses, however, the truth conditions are different for the two readings. For (6b), the deictic and coreference readings show strict identity (cf. §2.6): the deictic reading takes the two pronouns to refer to a specific individual not mentioned in the sentence, perhaps to Sister Bear, while the coreference reading takes them to refer to Papa Bear, with the second pronoun (in the ellipsed VP) linked to the overt pronoun in both cases. The bound-variable reading, on the other hand, shows sloppy identity. The pronoun is bound in both clauses but by different operators, so it refers to different individuals in each clause.

When we consider the three principles of the Binding Theory, Principles A and C look quite straightforward, and it is easy to see how children learn what an anaphor is and what is a name. Principle B is different and there is substantial learning involved, which seems to require parsing on the part of our children; the literature shows children having difficulty with Principle B. Elbourne 2005 surveys experimental work by many researchers investigating children's use of pronouns and the different ways they link to other DPs. If Principle B were simply part of the toolbox made available by UG like Principles A and C, we would expect similarly uniform linguistic behavior of children and rapid, accurate learning. Instead, we see children behaving quite differently, depending on the language they are selecting and their age.[1] Children appear to be challenged by the behavior of pronouns and to be conducting detailed analysis, sometimes arriving at systems that differ from those of the adults around them.[2]

We see that the Binding Theory needs to be part of the conceptual-intentional interface, part of what is given by UG, but it interacts with variable properties, which have to be learned and therefore involve parsing. No proposal for this kind of phenomena has been provided in terms of binary parameters whose content is stated at UG, as far as I am aware. If children are born to parse and if parsing is of fundamental importance, we can see how they might arrive at an appropriate analysis, incorporating much variation depending on the language being selected and the age at which learning takes place.

4.2   Phonological Form: Not Pronounced

Now let us turn to the sensorimotor interface and the requirements for the phonological form of expressions (one of the possible externalizations). We will be concerned with when elements may go unpronounced. Ellipsed VPs (VPs rendered silent through the operation of ellipsis) are unusual across languages, but English allows them, and children have plenty of evidence to that effect, as illustrated in (7). They occur in a wide range of structures, but they need a "host," an adjacent overt head that licenses them. The suggestion is that empty VPs occur only where they cliticize onto an adjacent host.[3] In (7a) the empty VP is the complement of *did*, and *did* hosts it. Of course, VP ellipsis only applies to VPs: (7b) is ill-formed because part of the VP remains, *for Naples*, and there is no null VP. In the ungrammatical (ii) structures of (7c,d), the null VP is separated from its potential host *had*, hence their ungrammaticality can be attributed to failure to cliticize. A properly hosted ellipsed VP may occur in a subordinate clause (7e), to the left of its antecedent (7f), in a separate sentence from its antecedent (7g), within a complex DP (7h), with an antecedent that is contained in a relative clause (7i), or even without any overt antecedent (7j).

(7) a. Max left on Wednesday but Mary did ~~$_{VP}$[leave on Wednesday]~~ as well.
   b. *Max left for Rio but Mary didn't $_{VP}$[~~leave~~ for Naples].
   c. i. They denied reading it, although they all had $_{VP}$[~~read it~~].
      ii. *They denied reading it, although they had all $_{VP}$[~~read it~~].
   d. i. They denied reading it, although they often/certainly had $_{VP}$[~~read it~~].
      ii. *They denied reading it, although they had often/certainly $_{VP}$[~~read it~~].
   e. Max left for Rio, although Mary didn't $_{VP}$[~~leave for Rio~~].
   f. Although Max couldn't $_{VP}$[~~leave for Rio~~], Mary was able to leave for Rio.
   g. Susan went to Rio.
       Yes, but Jane didn't $_{VP}$[~~go to Rio~~].
   h. The man who speaks French knows $_{DP}$[the woman who doesn't $_{VP}$[~~speak French~~]].
   i. People who appear to support mavericks generally don't $_{VP}$[~~support mavericks~~].
   j. Don't $_{VP}$[~~??~~]!

It appears that an ellipsed VP must cliticize (or incorporate in some other way) to the left, to a host head of which it is the complement:

(8) Max could visit Rio and Susan $_{INFL}$could + $_{VP}$[~~visit Rio~~], too.

This requirement explains the nonoccurrence of (9a), noted in Zagona 1988: the ellipsed VP needs an appropriate, adjacent host, a full phonological word, of which it is the complement, as in (9b). In (9a), *has* has become part of the noun *John* and no longer heads a phrase of which the empty VP is the complement.

(9) a. *I haven't seen that movie, but John's $_{VP}$[~~seen the movie~~].
   b. I haven't seen that movie, but John [has + $_{VP}$[~~seen the movie~~]].

Consider now null complementizers and deleted copies, where something similar seems to be at work, as discussed briefly in §1.1 (see (5)–(8) there). A child might hear sentences like (10a–c) pronounced with or without the complementizer *that*, because in English both versions occur. Such experiences would license an operation of the form in (10d) whereby *that* is deleted or rendered silent. French, Dutch, and German children have no comparable experiences and hence no grounds to parse a comparable deletion operation in their grammars; nothing like (10d) is triggered and there is no optionality for them; the complementizer must be present.

(10) a. Peter said [that/0 Kay had left already].
    b. The book [that/0 Kay wrote] arrived.
    c. It was obvious [that/0 Kay left].
    d. that → 0

So experience licenses the operation in (10d) for children acquiring English; but a linguist may observe that as a generalization, (10d) breaks down at certain points: *that* may not be null in the contexts of (11). The crucial data here are negative data, data about what does not occur, which are not available to children. Hence UG must be playing some role.

(11) a. Peter said yesterday [that/*0 Kay had left already].
    b. The book arrived yesterday [that/*0 Kay wrote].
    c. [that/*0 Kay left] was obvious to all of us.
    d. Fay believes, but Kay doesn't, [that/*0 Ray is smart].
    e. Fay said Ray left and Tim $_\vee e$ [that/*0 Jim stayed].
    f. Fay said [that/0 [that/*0 the moon is round] is obvious].

What we see here is that, much as with ellipsed VPs, *that* can be deleted only if the clause it occurs in is the complement of an overt, adjacent word. In (11a,b) the clause is the complement of *said* and *book* respectively, neither adjacent.[4] In (11c), the clause is the complement of nothing. In (11d) it is the complement of *believes*, which is not adjacent, and in (11e) it is the complement of a verb that is not overt. In (11f) the lower complementizer may not be null because its clause is not the complement of *said*.[5]

The same condition holds for what we used to view as "traces" of *wh-* movement. English-speaking children learn that *wh-* elements are displaced, that is, pronounced in a position other than where they are understood, on hearing and understanding a sentence like (12a). On Minimalism's Copy-and-Delete implementation of displacement, there are actually multiple copies of the same element; an independent principle says that only one of them may be pronounced (in this case, it is the sentence-initial one), entailing deletion of all the others. For more on this, see note 7. In (12a',b'), the structures posited for (12a,b), the lowest *who* is the complement of the adjacent verb, and in (12b'), the intermediate *who* occurs in a clause that is the complement of the adjacent verb *say*.

(12) a. Who did Jay see?
     b. Who did Jay say ~CP~[that Fay saw]?
        a'. Who did Jay see ~~who~~?
        b'. Who did Jay say ~CP~[~~who~~ that Fay saw ~~who~~]?

Assuming the Copy-and-Delete Minimalist structures of (12a',b'), a copy of *who* can be deleted only when it or the clause in which it occurs is the complement of an adjacent, overt word. If that is the condition, it predicts, with no further learning, that (13a) is ill-formed, because the boldface **who** is undeletable (henceforth, boldface indicates a copy that cannot be deleted as required): it is in a clause that is the complement of *apparent* but not adjacent to it. The lowest *who* is the complement of the adjacent, overt *seen*, hence deletable. Also, if *yesterday in Chicago* were not present in (13a), then it *would* be the case that *who* was in an adjacent complement of the overt *apparent*, hence deletable; this yields the well-formed (13b), where (13b') is the Copy-and-Delete representation.

(13) a. *Who was it apparent yesterday in Chicago $_{CP}$[**who** that [Kay had seen ~~who~~]]?
   i.e.,
   *Who was it apparent yesterday in Chicago who that Kay had seen?
   and
   *Who was it apparent yesterday in Chicago 0 that Kay had seen?
   b. Who was it apparent that Kay had seen?
   b'. Who was it apparent [~~who~~ that Kay had seen ~~who~~]?

We thus solve the poverty-of-stimulus problem posed by (13a) as follows: children learn simply that *wh-* items may be displaced (copied and deleted), and the interface condition requiring deleted items to cliticize onto an adjacent host causes the derivation of (13a) to crash with no further learning.

Other contexts likewise indicate that items may be deleted only if they are the complement or in the complement of an overt, adjacent word. So *which man* is deletable in the leftmost conjunct of (14c) (the complement of the adjacent *introduce*) but not the boldface **which woman** in the rightmost conjunct, the complement of a nonovert verb. Hence the corresponding sentence is ill-formed. Similarly, in (14d,e,g), the boldface element fails to meet the condition for deletion, because the relevant verb is not overt. These structures involve *wh-* movement (14c,d), readily learnable as noted above; heavy-DP shift (14e,g), learnable on exposure to simple expressions like *John gave to Ray his favorite racket*; and gapping (14c,d,e,g), learnable on exposure to things like (14b,f). The UG principle then solves the poverty-of-stimulus problems of (14c,d,e,g).[6]

(14) a. Jay introduced Kay to Ray and Jim introduced Kim to Tim.
   b. Jay introduced Kay to Ray and Jim $_{V}e$ Kim to Tim.

c. *Which man$_i$ did Jay introduce ~~which man$_i$~~ to Ray and
   which woman$_j$ Jim $_ve$ **which woman$_j$** to Tim?
   i.e.,
     *Which man did Jay introduce to Ray and which woman
     Jim which woman to Tim?
     and
     *Which man did Jay introduce to Ray and which woman
     Jim 0 to Tim?

d. *Jay wondered what$_i$ Kay gave ~~what$_i$~~ to Ray and what$_j$
   Jim $_ve$ **what$_j$** to Tim.

e. *Jay admired [~~his uncle from Paramus~~]$_i$ greatly [his uncle
   from Paramus]$_i$ but Jim $_ve$ **[his uncle from New York]$_j$**
   only moderately [his uncle from New York]$_j$.

f. Jay gave his favorite racket to Ray and Jim $_ve$ his
   favorite plant to Tim.

g. *Jay gave [~~his favorite racket~~]$_i$ to Ray [his favorite racket]$_i$
   and Jim $_ve$ **[his favorite plant]$_j$** to Tim [his favorite
   plant]$_j$.

The same condition explains why a complementizer may not be
null if it occurs to the right of a gapped (nonovert) verb, as in (15b);
nor does one find a deleted copy in that same position, as with the
boldface **who** in (15c).

(15) a. Jay thought Kay hit Ray and Jim $_ve$ $_{CP}$[that Kim hit Tim].
     b. *Jay thought Kay hit Ray and Jim $_ve$ $_{CP}$[0 Kim hit Tim].
     c. *Who$_i$ did Jay think Kay hit ~~who$_i$~~ and who$_j$ Jim $_ve$
        $_{CP}$[**who$_j$** (that) [Kim hit ~~who$_j$~~]]?

So, children exposed to some form of English have plenty of evidence that a *that* complementizer is deletable (10d), that *wh-* phrases
may be displaced (copied), and that heavy DPs may be copied to
the end of a clause (14e,g); but they also know *without evidence* that
complementizers and copies may not be deleted unless they are the
complement or in the complement of an adjacent, overt word. And

the data of (10–15) suggest that this is the information that UG needs to provide and that head–complement relations are crucial. The convergence of that information with the I-language-specific devices that delete a *that* complementizer and allow a *wh-* phrase or a heavy DP to be copied yields the distinctions we have noted and solves the poverty-of-stimulus problems.[7] The UG requirement guarantees that deleted items must be understood in structurally prominent positions, where they have an appropriate host. This might be motivated by parsing needs: the possibility of a deleted item need only be considered where there is an appropriate host for one. The absence of an appropriate host rules out a deleted element; correspondingly, the presence of an appropriate host is a potential cue to the presence of a deleted item.

More evidence for this interface requirement comes from failures of verb reduction. The verbs *is, am, are, has, have, had, will, would,* and *shall* may reduce: *Kim's happy, Jim'll do it, Sarah'd read it,* and so on. However, by now readers are not surprised that there are apparent exceptions: for example the boldface instances of *is* in *Kim's happier than Tim* **is**, *I wonder what the problem* **is**, *I wonder what that* **is** *up there, I wonder where the concert* **is** *on Wednesday* may not reduce. These data, negative data concerning contexts where *is* does not reduce, are not available to children directly, and that is the familiar poverty-of-stimulus problem: the stimulus appears to be too poor to determine all the properties of the mature system. Children hear some instances of the reduced forms but somehow come to know much more, namely that *is* may be reduced generally but not in the boldface contexts above. But notice that the boldface items each precedes a deletion site, as shown in (16). Our emerging analysis suffices to explain the nonreduction: the full form is needed to license the deletion site.[8]

(16) a. Kim is happier than Tim **is** ~~happy~~.

    b. I wonder what the problem **is** ~~what~~. (Cf. *The problem's twofold.*)

c. I wonder what that **is** ~~what~~ up there. (Cf. *That's a fan up there.*)

d. I wonder where the concert **is** ~~where~~ on Wednesday. (Cf. *The concert's in Nogales on Wednesday.*)

All is well so far, but now the question is: *how* is *what* deleted? Let us first review effects of earlier restrictions and see how we might capture them with the economy and elegance that the Minimalist Program encourages.

We know that elements may cliticize to the left and become an inseparable part of their host. That happens with the reduced *is* discussed earlier. When *is* reduces, its pronunciation is determined by the last segment of the word to which it attaches, as (17a) illustrates: voiceless if the last segment is voiceless, voiced if the last segment is voiced, and syllabic if the last segment is a sibilant or affricate. Precisely the same is true of the plural marker, the possessive, and the third-person singular ending on a verb, illustrated in (17b–d) respectively.

(17) a. Pat's happy, Doug's happy, and Alice's here.

b. cats, dogs, and chalices

c. Pat's dog, Doug's cat, and Alice's crocodile

d. commits, digs, and misses

Children understand *Pat's happy* as 'Pat is happy', *Pat* being the subject of the phrase 'is happy'. However, *is* is pronounced inseparably with *Pat*, and children parse what they hear as (18a), that is, with reduced *is* attached to the noun, with normal pronunciation applying. What (18a) expresses is a piece of structure, (18b), that serves to determine the shape of the emerging grammar, showing particularly that elements may be cliticized (Lightfoot 1999, 2006a). So from hearing and understanding an expression like *Pat's happy*, children learn that *is* may be reduced and absorbed into the preceding word. Again we see the effects of parsing.

(18) a. $_N$Pat + 's
    b. noun + clitic

If we draw (17) together with (19), we now find something interesting: copies do not delete if they are to the right of a cliticized verb. In (19), the copied elements may be deleted if *is* is in its full form, but not if it is reduced; the corresponding sentences with *'s* do not occur.

(19) a. Kim is happier$_i$ than Tim is/*Tim's ~~happy~~$_i$.
    b. That is a fan up there.
    c. I wonder what$_i$ that is/*that's ~~what~~$_i$ up there.
    d. I wonder where$_i$ the concert is/*concert's ~~where~~$_i$ on Wednesday.

This suggests again that a deleted copy is incorporated into the element of which it is the complement. In (19), if *is* cliticizes onto the subject noun and becomes part of that noun, it no longer heads a phrase of which *what/where* is the complement, and no incorporation is possible, hence no deletion if deletion is incorporation or cliticization.

That idea enables us to capture another subtle and interesting distinction. The sentence in (20a) is ambiguous: it may mean that Mary is dancing in New York or just that she is in New York (working on Wall Street, say, not dancing). The minimally different (20b), however, only has the latter interpretation. The 'dancing in New York' interpretation of (20a) has a structure with an empty verb, understood as 'dancing', represented in (20c). If empty elements (like an understood verb) are incorporated, there must be an appropriate host. There is an appropriate host in (20c), where the empty verb cliticizes onto a full verb, *is*, but not in (20d): $_V e$ isn't the complement of *Mary's*, therefore it is not licensed. Consequently (20b) unambiguously means that Mary is in New York (occupation unspecified), because there is no empty, understood verb. Again, it is inconceivable that children *learn* such distinctions purely on the basis of external evidence.

(20) a. Max is dancing in London and Mary is in New York.
     b. Max is dancing in London and Mary's in New York.
     c. Max is dancing in London and Mary is $_\vee e$ in New York.
     d. *Max is dancing in London and Mary's $_\vee e$ in New York.

So copies are deleted in the phonology in order to satisfy linearization requirements, and our analysis takes deletion to be an instance of cliticization, which allows the analysis to generalize to other null elements, such as copies, as already discussed above. In (21a) the deleted complement cliticizes onto the adjacent *see*, and in (21b) the deleted *Jay* is in the complement of *expected*, which is adjacent to it, and accordingly cliticizes onto it.

(21) a. Who$_i$ did Jay see ~~who~~$_i$?
     b. Jay$_i$ was expected [~~Jay~~$_i$ to win].

The analysis appeals to head–complement relations and adjacency.

Our analysis captures many other distinctions. For example, English speakers' grammars typically have an operation whereby a "heavy" DP is displaced to the right (see (14e,g) above). Under our Copy-and-Delete approach that means merging a copy to the right and reducing the first copy to silence by absorbing it clitic-like into a host. In (22a) the copied element is the complement of *introduced*, hence incorporated and deleted successfully; in (22b) it is in the complement of the adjacent *expect*; but in (22c) the element that needs to be deleted is neither the complement nor contained in the complement of anything, and the derivation is ill-formed and crashes.

(22) a. I introduced [~~all the students from Brazil~~]$_i$ to Mary [all the students from Brazil]$_i$.
     b. I expect [[~~all the students from Brazil~~]$_i$ to be at the party] [all the students from Brazil]$_i$.

    c. *[[**All the students from Brazil**]$_i$ are unhappy] [all the
       students from Brazil]$_i$.

Our UG principle, that deletion of this kind is cliticization or incorporation, solves the poverty-of-stimulus problem of (22c): children
simply learn that heavy DPs may be copied to the right, and the UG
condition accounts for the nonoccurrence of (22c) with no further
learning or experience needed.

    Our analysis can also solve a puzzle about genitives and DP
structure, discussed in §3.1. Whereas a simple DP like *a book* has the
structure $_{DP}$[$_D a$ $_N book$], a DP like *Kim's book about syntax* has
the Determiner *'s* governing (and assigning Case to) its specifier,
the genitive *Kim*, as well as its complement $_{NP}$[*book about syntax*].
Consider now an expression like *Jay's picture*. It is three-ways
ambiguous: Jay may be the owner of the picture, the painter, or the
person portrayed. The latter reading is the so-called objective genitive and is usually analyzed as in (23), where *Jay* is copied from
the "object" position to the specifier of the DP. The operation is specific to grammars of English speakers and does not occur in
French, for example. This much is learnable: children hear expressions like *Jay's picture* in contexts where it is clear that Jay is
pictured.

(23) $_{DP}$[Jay$_i$'s $_{NP}$[picture ~~Jay$_i$~~]]

The curious thing is that comparable expressions like *the picture
of Jay's, The picture is Jay's,* and *the picture that is Jay's* show only
a two-way ambiguity, where Jay may be the owner or the painter
but not the person portrayed. This is yet another poverty-of-stimulus
problem, because it is inconceivable that children are systematically
supplied with evidence that the objective interpretation is not available in these cases. We have an explanation for this, as already
noted in §3.1: the structure of these expressions would need to be
as follows.

(24) a. *the picture of $_{DP}$[Jay's $_{NP}$[~~picture~~ **Jay**]]   (*the picture
            of Jay's*)
     b. *the picture is $_{DP}$[Jay's $_{NP}$[~~picture~~ **Jay**]]   (*the picture
            is Jay's*)
     c. *the picture that is $_{DP}$[Jay's $_{NP}$[~~picture~~ **Jay**]]   (*the picture
            that is Jay's*)

A preposition like *of* in (24a) is always followed by a DP, a possessive like *Jay's* occurs only as the fused specifier and head of a DP,
and Ds always have an NP complement, even if the noun is empty,
as it is here (where it is understood as 'picture'). Now we can see
why the structures are ill-formed: the copied *Jay* has no host to cliticize onto, hence it is undeletable (boldface) and the derivation
crashes. *Jay* is the complement of the adjacent noun, but that noun
is not overt, hence not a viable host.

The pair in (25) reflects another distinction covered by our
account. The sentence in (25a) is well-formed and involves no
deletion of a copied element, whereas (25b) involves two instances
of DP copying and deletion (to yield the passive constructions).
The leftmost instance is well-formed, because the copied *Jay* is in
the complement of the adjacent *known* and therefore deletes; however, in the rightmost conjunct, the copied *he* has no overt host to
cliticize onto and therefore cannot be deleted as required, leading
the derivation to crash.

(25) a. It is known that Jay left but it isn't $_V e$ that he went to the
            movies.
     b. *Jay$_i$ is known [~~Jay$_i$~~ to have left] but he$_i$ isn't $_V e$ [**he**$_i$ to
            have gone to the movies].

And there is more: it is well known that an expression like *They
were too angry to hold the meeting* is ambiguous, meaning either
that they were so angry that they couldn't hold the meeting or that
some unspecified person (e.g., the speaker) couldn't hold the meeting; the ambiguity lies in who was in charge of holding the meet-

ing (Chomsky 1986: 33). The former reading has the structure of (26a), where *they* is copied and deleted; the CP is the complement of *angry* and *they* is in that complement and adjacent to *angry*, hence incorporated. The other reading has arbitrary PRO as the subject of *hold*, as shown in (26b): nothing is copied, and that would not be possible because the clause is an ADJUNCT to *angry*, not a complement (adjuncthood is represented here by italics).

(26) a. They$_i$ were too angry $_{CP}$[~~they~~$_i$ to hold the meeting].
   b. They were too angry $_{CP}$*[PRO$_{arb}$ to hold the meeting]*.
   c. Which meeting$_i$ were they$_j$ too angry $_{CP}$[~~which meeting~~$_i$ [~~they~~$_j$ to hold ~~which meeting~~$_i$]]?
   d. *Which meeting$_i$ were they$_j$ too angry $_{CP}$*[**which meeting**$_i$ [PRO$_{arb}$ to hold ~~which meeting~~$_i$]]*?

However, the corresponding question *Which meeting were they too angry to hold?* is unambiguous and has only the anaphoric reading, as in (26c), under which *they* are unable to hold the meeting. It lacks the meaning of an arbitrary subject for *hold*: (26d) is ill-formed. In (26c), the clause is the complement of *angry* and therefore *which meeting* in that complement can cliticize onto *angry* and thus be deleted. Likewise, the copied *they* is deleted successively in (26c). (See also (31).) However, in (26d), the clause is an adjunct to *angry*, not a complement, and therefore the intermediate copy of *which meeting* is undeletable.

Several instances of deletion, we have now seen, are subject to poverty-of-stimulus problems suggesting a cliticization or incorporation analysis. Our children are learning what they need to learn through parsing positive data. Other instances of apparent deletion are not subject to comparable poverty-of-stimulus problems and do not fall under a cliticization treatment. Van Craenenbroeck and Merchant 2013 offer a quite comprehensive inventory of deletion processes, instances where elements are not pronounced. In some instances we understand analyses in some detail, but other examples are less well understood, and work remains to be done on why a cliticization or incorporation analysis works in some places and not elsewhere. Nonetheless the poverty-of-stimulus problems are real and require at least the information invoked here, even if analyses require further elaboration. For example, gapped verbs have a very different distribution from ellipsed VPs, so they do not cliticize in the way that we have analyzed ellipsed VPs here. Compare the gapped verbs in (14, 27) with the ellipsed VPs in (7): their distribution is quite different.

(27) a. *Max speaks French, although Mary $_\text{V}e$ German.
     b. *Jim said that Max speaks French and Kim said that Mary $_\text{V}e$ German.
     c. *Max $_\text{V}e$ French and Mary speaks German.
     d. *The man who speaks French knows $_\text{DP}$[the woman who $_\text{V}e$ German].
     e. *Max drove to New York and Susan did $_\text{V}e$ to Chicago.[9]

So far we have been talking about deletion sites as involving cliticization onto a host, treating the deleted item as some kind of clitic. Indeed, it is profitable to view the incorporated items as clitics. Zwicky and Pullum (1983) distinguish between clitics and AFFIXES, and this distinction permits some further understanding. Specifically, Zwicky and Pullum argue that the English reduced negative *n't* is an affix: so in our terms *isn't*, for example, is formed in the lexicon and merged directly into syntactic structure. That distinguishes between (28b), where *isn't* is merged with *here* to form a constituent, and the ill-formed (28c).

(28) a. John's not here.
     b. John isn't here.
     c. *John'sn't here.

Two of Zwicky and Pullum's criteria for their distinction are given in (29). Criterion F says that affixes may not attach to material already containing clitics, hence the nonoccurrence of (28c).

(29) E. Syntactic rules can affect affixed words, but cannot affect
         clitic groups.
     F. Clitics can attach to material already containing clitics,
         but affixes cannot.

This allows us to distinguish between the structures of (30): criterion E allows a syntactic copying operation (what we used to think
of as displacement or movement) to affect *couldn't*, an affixed form,
but not *could've*, where *'ve* is cliticized onto *could*.

(30) a. Couldn't Kim see that?
     b. *Could've Kim seen that?

Hence also the grammaticality of the corresponding *Could Kim've
seen that?* versus *Could Kimn't see that?*

If *n't* is an affix, then phonologically reduced verbs (*'s, 've, wanna,*
etc.), ellipsed VPs, null complementizers, gapped verbs, and deleted
copies are clitics. If clitics may attach to material already containing clitics (29F), we allow (31a–d) but not (31e), which has an affix
attached to *could've*, in violation of (29F).[10]

(31) a. Kim visited NY and Jim could've $_{VP}e$.
     b. Kim visited NY but Jim couldn't $_{VP}e$.
     c. Kim visited NY but Jim couldn't've $_{VP}e$.
     d. I'd've visited NY.
     e. *Jim could'ven't seen it.

There is a vast literature on clitics and many distinctions are
drawn; indeed, Arnold Zwicky argued in his later work that there
are no clitics (Zwicky 1994). I have drawn selectively from that literature in arguing that the deletion sites discussed so far are clitics. However, it may be that the incorporation analysis of deletion
is correct but that the incorporated elements are not clitics; the
claims are logically distinct. Thinking of the deletion of copied
phrases as cliticization enables us to understand old puzzles about
the Fixed-Subject Condition (Bresnan 1972) and the *that*–trace
effect of the 1970s, later subsumed under the agreement relations

of Rizzi 1990. It also enables us to learn more about the cliticization operation. In general, subjects resist displacement; when they are copied into a displaced position, odd things happen (for discussion, see Lightfoot 2006b).[11]

Not only do complementizers like *that* and *how* not generally host clitics (see note 11), neither do prepositions. This explains the well-known observation that generally prepositions do not license movement sites: French **Qui as-tu parlé avec?*, Dutch **Wie heb je met gesproken?*, 'Who have you spoken with?'. In English, prepositions may be stranded like this, but only where they are themselves reanalyzed as part of a complex verb, as in (32a) (see Hornstein & Weinberg 1981 for discussion of the reanalysis operation); compare the ill-formed (32b,c), where the PP is not the complement of an adjacent verb (in (32b) it is not adjacent, in (32c) it is an adjunct) and consequently may not host the deleted copy.

(32) a.   Who$_i$ did you $_v$talk + to ~~who$_i$~~?
     b.   *Who$_i$ did you talk at the meeting to **who$_i$**?
     c.   *What$_i$ did you sleep during **what$_i$**?

I have argued that English speakers *learn* that certain verbs may be phonologically reduced, that complementizers may be null, that *wh*- phrases may be displaced (pronounced in positions other than where they are understood), that verbs may be gapped, that heavy DPs may be displaced to the right, that VPs may be ellipsed, that possessive noun phrases may have objective interpretations. These seven variable properties are readily learnable from the linguistic environment, and we can point to plausible PLD. Such data that all English-speaking children hear include sentences like *Kim's happy*, manifesting reduction; *Peter said Kay had left already* (11a), exhibiting a null complementizer; *Who did Jay see?* (12a), with a displaced *wh*- phrase; *Jay introduced Kay to Ray and Jim Kim to Tim* (14b), an example of gapping; *Jay gave to Ray his favorite racket* (14g), heavy-DP shift; *Max could visit Rio and Susan could, too* (8), an ellipsed VP; and *Jay's picture* (23), meaning 'picture of Jay'.

The way to think of this, I believe, is that children identify certain structures, through understanding and assigning structure to what they experience, that is, through parsing; some of these structures reflect variable properties. Consider the object–verb-order parameter. If we take parsing to be the key, children find either $_{VP}$[DP V] or $_{VP}$[V DP] structures, very specific information. Children use structures or lose them: a child who builds object–verb $_{VP}$[DP V] into her I-language loses $_{VP}$[V DP] structures, which atrophy. Notice that children are reacting to abstract structures, elements of grammar, which are required to understand expressions that they hear; they identify only structures that are unambiguous.

I have argued that an empty element (a deleted phrasal copy, a null complementizer, an ellipsed VP, the ellipsed *dancing* in 20b,c) is incorporated or cliticized onto an adjacent phonological head (N, V, Infl) of which it is (in) the complement. This one simple idea at the level of UG interacts with seven grammar-specific devices, all demonstrably learnable, and that interaction yields a complex range of phenomena. This involves carving up the grammatical world differently.

We seek a single object: the genetically prescribed properties of the language organ. Those properties permit language acquisition to take place in the way that it does, and that means that we must examine language variation along the lines of Baker 2001; that yields a wealth of empirical considerations. Baker analogized parametric options in language to the *elements* of chemistry, claiming that the linguistic options are the basic building blocks of languages. That imputes much detailed information to UG in violation of Minimalist principles. What we postulate must solve the poverty-of-stimulus problems that we identify and solve them for *all* languages as well. We also want our ideas to be as elegant and economical as is feasible. In addition, the grammars that our theory of UG permits must meet other demands.

To take just one example, they must allow speech comprehension to take place in the way that it does. That means that considerations of parsing might drive proposals. That hasn't happened much yet, but there is no principled reason why not, and the situation might change. Similarly, evidence drawn from brain imaging or even from brain damage might suggest grammatical properties. In fact, the proposals here look promising for studies of online parsing. When a person hears a displaced element, say a *wh-* phrase at the beginning of an expression, she needs to search for the deletion site, the position in which it needs to be understood. The ideas developed here restrict the places where she can look.

Here I have tried to sketch the details of what a good theory of parsing would lead a child to select. We are far from a satisfactory theory, but thinking in terms of how children interpret the contrasts they experience looks far more tractable than seeking to define UG-defined parameters of what constitutes what kind of clitic. The latter would entail postulating very rich information as part of UG, violating Minimalist aspirations.

One uses what looks like the best evidence available at any given time, but that will vary as research progresses, and consequently the form of our innateness claims will vary. There are many basic requirements that our hypotheses must meet, and there is no shortage of empirical constraints, and therefore there are many angles one may take on what we aim for. In this chapter I have taken one angle and progressed beyond where government took us: to delete an element is to cliticize it. This is certainly not the end of any story, but a reasonable way to proceed and an improvement on earlier accounts.

5 Population Biology: The Spread of New Variable Properties

5.1 Discontinuities and Abstractions

If syntacticians walk the four hundred miles from Berlin to Amsterdam, covering ten miles a day, it is said that they will hear no noticeable difference in the language of local people they meet at breakfast and at dinner.[1] Not even on the day when they cross the border from Germany into the Netherlands. The paradox is that the German of Berlin is indeed different from the Dutch of Amsterdam.

As in space, so in time. The messages that my daughters send are similar to the letters my mother used to send. The syntax is the same and the vocabulary close to identical, though they might use a few different words with different meanings. But a similar paradox applies: local differences are slight, but if we look over longer time spans, we see big differences across the language of Geoffrey Chaucer, William Shakespeare, Jane Austen, and Toni Morrison.

In many contexts, between neighboring towns and villages or between generations of a family, language seems to be stable and its transmission frequently seems close to perfect. Change is often gradual to the point of being imperceptible, but when we use a longer lens, we see major discontinuities.

There are many understandings of this paradox, depending on one's broad view of what drives syntactic change and acquisition. Nineteenth-century neogrammarians developed a theory of sound change but nothing parallel for syntax, morphology, or other aspects of language. This was their legacy for American structuralists, who also limited their work to what we now call phonology and to narrowly descriptive work in syntax and morphology.

A typical, linguistics-textbook treatment from the 1950s depicts language change as change in sounds and, crucially for our discussion here, as necessarily gradual: "we described sound change as a gradual change in habits of articulation and hearing, taking place constantly, but so slowly that no single individual would ever be aware that he might be passing on a manner of pronunciation different from that which he acquired as a child. This gradualness is extremely important" (Hockett 1958: 439). Hockett goes on, "when a person speaks, he aims his articulatory motions more or less accurately at one after another of a set of bull's-eyes, the allophones of the language" (p. 440). Speakers are "quite sloppy in [their] aims most of the time" and over time may hit different targets, hence language change. Thus, for many in the 1950s, language change was gradual change in sound production.

Kiparsky 1968: 175 describes this view as tantamount to seeing a language as a "gradual and imperceptibly changing object which smoothly floats through time and space," changing, for example, from Old, to Middle, to Early Modern, and then to Present-Day English, with various gradations in between but no discontinuities or major disruptions. That view is still often expressed, as we shall see.

Kiparsky and his colleagues, on the other hand, were early generativists dealing with change. They worked with abstract grammars encompassing syntax and morphology alongside sounds and viewed them as changing when children encountered new ambient language. An abstract grammar was a system with its own structures and computational operations, a biological object selected by children and represented in people's brains. For Kiparsky, "the transmission of language is discontinuous, and a language is recreated by each child on the basis of the speech data it hears." If one abstract structure in a generative system changes, that typically has multiple consequences for the structures and expressions generated by the system, leading to discontinuities and possibilities for large-scale changes.

Although change is often gradual to the point of being imperceptible, the major discontinuities that arise stem from the abstractness of the structures. Lightfoot 1979 and 2017c reiterates Kiparsky's view that grammars are invented or selected by children, not transmitted wholesale to the next generation, and summarizes arguments for some major and now well-understood discontinuities. I show how the paradox of the syntactician walking through imperceptible variations from Berlin to Amsterdam and ending up in a different language might be explained by a nonstandard approach to language acquisition that employs a clear distinction between sociologically defined external language and the internal languages of minds/brains. I argue that children discover and select the abstract structures of their internal languages. New phenomena in the input may trigger a single change at the abstract level, yielding a new structure that serves to generate many new phenomena that enter the language at the same time.

This approach to language change anticipates neo-Darwinian biologists' appeal to punctuated equilibrium, in turn based on evolutionary biologist Ernst Mayr's model of geographic speciation (Mayr 1942; Eldridge & Gould 1972), and their rejection of Darwin's gradualism. In many disciplines, this focused attention on structural shifts, known as "catastrophes" (Thom 1972)[2] or "phase transitions" at different stages of investigation within the successive frameworks of catastrophe theory in the 1950s and 1960s; chaos theory (Gleick 1987) and synergetics (Haken 1984) in the

1970s and '80s; and complexity science more recently (Casti 1994; Kauffman 1995; Prigogine & Stengers 1997).

Linguists have identified such saltations and understood them in terms of children selecting new systems when exposed to new ambient language. Sometimes several phenomena change simultaneously, a catastrophe or phase transition, and one can explain the simultaneity by arguing that there is a single change at the abstract level of the internal system from which the new phenomena follow. The task then is to show how the new system might have been discovered and selected by children; how new ambient language (i.e., new external language) triggers new internal systems. Rich and deep explanations have been developed for some syntactic changes, as outlined in §2.4–§2.6 and §3.3–§3.5. These explanations were deepened by a conceptual shift in the 1980s, which might be characterized as the great individualization.

5.2   Individualism

Noam Chomsky famously defined his field as "concerned primarily with an ideal speaker-listener, in a completely homogeneous speech-community, who knows its language perfectly and is unaffected by such grammatically irrelevant conditions as memory limitations, distractions, shifts of attention and interest, and errors (random or characteristic) in applying his knowledge of the language in actual performance" (1965: 3). Under this idealization, linguists wrote grammars for complete languages like Finnish, English, Japanese, and so on, capturing the competence of idealized speaker-listeners.

However, a subsequent work, Chomsky 1986, elaborated this competence–performance idealization and introduced a distinction between external language and internal languages, very different kinds of things. As discussed in §2.2, E-language is a mass-sociological concept, a group phenomenon, external language out there, the kind of unanalyzed thing that anybody might hear, and it includes the PLD that trigger new systems. I-languages, on the other hand, are internal, individual systems that emerge in children according to the dictates of the inherent language capacity and to the demands of the ambient E-language to which they are exposed and which they seek to understand and analyze, that is, to parse. I-languages are properties of individual minds/brains and consist of abstract structures. Individuals acquire an individual, private I-language, creating their particular form of English, perhaps the language of a seventy-five-year-old woman in Padstow in northern Cornwall, and not the external English language as a whole or even the English of her community.

Chomsky now echoed Wilhelm von Humboldt, who distinguished the language of individual citizens from the language of a nation ([1836] 1971), and Hermann Paul, who asserted that "die reelle Sprache nur im Individuum existiert" (real language exists only in individuals; 1877: 325) and later that "wir müssen eigentlich so viele Sprachen unterscheiden als es Individuen gibt" (we must in fact distinguish as many languages as there are individuals; 1880: 31).

Put differently, "English" is not recursively enumerable. There is no system that will generate the sentences of English, partly because of internal contradictions. For example, an expression like *Kim might could read this* is an expression of English in Arkansas but not in New York or Cornwall. *Gwain ee t'Exeter* 'is he going to Exeter?' is said in Cornwall, perhaps by our seventy-five-year-old Padstow woman, but not in New York or Arkansas. Under this view, the English language has no more reality than the French liver, English irony, or the Scottish love of whisky; such things do not exist except as abstract idealizations.

Similarly, and relevantly for the catastrophe/phase transition to be discussed in §5.4, there is no recursive device that generates the set of expressions found in surviving English texts from the thirteenth and fourteenth centuries. Rather there is a *set* of I-languages that generates what we see in the texts. Thus, we study language change in part as the spread of new I-languages, using the methods of population biology (see §5.9). There is no biologically coherent notion of English, certainly not as an object being transmitted steadily from one generation of language users to another. If languages are not transmitted, there are many damaging consequences for traditional ways of thinking about language change, but that is a topic for other days.

Chomsky made his move toward I-languages at about the same time that Tony Kroch introduced his idea of competing, coexisting grammars: individuals often have more than one internal system and may use different systems at different times in a kind of internal multiglossia (1989; see also Yang & Roeper 2011). If individuals may use more than one I-language, then we may go beyond Paul and argue that there are many more I-languages than there are individuals. Bear in mind that, under the approach we have adopted here, a person's individual I-language is the means by which she parses what she hears, so she may need multiple I-languages to deal with the diversity of the ambient language. It may not be possible to parse what one hears from visiting relatives of different ages and from different countries using a single I-language.

These ideas have had profound consequences for syntactic analyses and, in particular, for thinking about syntactic change, consequences that have still not been fully thought through. They enable us to gain new understanding of catastrophes or phase transitions, when many phenomena change at the same time, or of domino effects, when changes occur in rapid succession. The ideas of E-language and I-languages suffice to account for language acquisition, without the conventional, sociologically defined notion of English or Estonian. With these notions, linguists do not write grammars for whole languages but for I-languages that capture an individual's linguistic capacity. One views children as selecting their private I-language on exposure to E-language. Children parse the E-language they hear and acquire the categories and structures needed to understand what they hear. Thus they discover and select the elements of their I-language (Lightfoot 2017b). A person's I-language is the aggregation of the structures permitted by UG and those required to understand what she hears (i.e., those already discovered and selected).

In §5.3 I will sketch some views about the relationship between major discontinuities in I-languages, catastrophes or phase transitions, and what writers regard as unusual events. In §5.4 I will outline the I-language–E-language distinction in greater detail and sketch how our discovery-and-invention/selection approach to acquisition enables us to understand the emergence of new I-languages with multiple new variable properties, specifically in the remarkable development of Middle English. In §5.5–§5.6, I will outline again our particular discovery approach to acquisition and consider through that lens the central paradox that we defined at the beginning of this chapter. In §5.7, we will work through some details of discontinuities first discussed in chapter 2, by considering how children were driven to the discontinuities revealed through the new internal systems, the new I-languages that children grew. In §5.8, we will think about the spread of new I-languages, and in §5.9, we will use the methods of population biology to explore phase transitions in language and will discuss a computer simulation.

5.3 Discontinuities and Unusual Events

The idea that language transmission generally takes place more or less perfectly and that children acquire the same language capacity as their neighbors and parents (unless something unusual happens) is still with us. For example, Jürgen Meisel has argued (2011) that change takes place only in multilingual contexts, which he takes to involve sociologically distinct languages like German and Turkish. Discontinuities are viewed as unusual events provoked by particular social contexts, as when people are exposed to multiple languages, in the conventional sense of the term. So in recent generations, the multilingualism of Turks and Germans in Germany has had effects on the two languages.

However, multilingualism in Meisel's sense does not explain how languages like Icelandic, which have been isolated for a long time, nonetheless undergo changes in their syntactic structures, including major shifts in word order. Furthermore, if one works with conventional definitions of languages as properties of groups, one underestimates the variation and multilingualism found everywhere (Lightfoot 2011). There is significant variation among people within the same speech community, indicating individuals have their own private systems. In §5.5, we will reject the conventional notion of a socially defined language as the locus of change and distinguish internal and external languages, arguing that real change is best defined at the level of individual language users (Lightfoot 1993).

A similar predilection for stasis in normal times, with change only as a consequence of unusual events, comes from recent ideas about "inertia," developed quite differently in Longobardi 2001 and Keenan 2002 (see Roberts 2017 for good discussion of the differences). The central idea is that "things stay as they are unless acted upon by an outside force or decay" (Keenan 2002: 327) or that "syntactic change should not arise, unless it can be shown to be *caused*—that is, to be a well-motivated consequence of other types of change (phonological changes and semantic changes, including the appearance/disappearance of whole lexical items) or, recursively, of other syntactic changes" (Longobardi 2001: 278). These ideas were developed in part as a reaction against the notion that changes may be internally motivated by "UG biases" and cyclical forces (Roberts & Roussou 2003; Van Gelderen 2011). Inertia approaches require an external cause for change, but the external cause might be the result of a prior change in internal grammars (for critical discussion, see Walkden 2012).

Another view is that sometimes there is "imperfect learning" (Kiparsky 1968; Trudgill 2002; Mitchener & Nowak 2004; Montrul 2008): children are exposed to the same linguistic experience as their parents, but their learning is imperfect and they converge on a different mature system, yielding imperfect transmission, and hence discontinuity. Imperfect learning may be a special event, but it is not triggered by new experiences and can take place at any time, yielding discontinuities randomly.

5.4   The Case of Anglicized Norse

Emonds and Faarlund (2014) offer a radical challenge to the philologists' conception of English progressing gradually and imperceptibly from one stage to another. The authors, henceforth E&F, postulate that so-called early Middle English—spoken and, crucially for historical linguists, *written* in the East Midlands in the twelfth and thirteenth centuries—reflects the results of intensive and extensive contact and represents a new language, which they call Anglicized Norse. It has many features of Scandinavian syntax alongside West Germanic words and phonological structures corresponding to Old English antecedents. They argue that Anglicized Norse syntax essentially replaced that of Old English. E&F build on work by philologists and syntacticians who argue for analyses that E&F construe now as aspects of a more comprehensive phase transition, and they muster considerable evidence for their analysis. However, their work has had, it is fair to say, a hostile reception from many historical linguists, notably in an issue of *Language Dynamics and Change* devoted to discussion of their book by several authors (e.g., Holmberg 2016; McWhorter 2016; Trudgill 2016) and in a review article by Kristin Bech and George Walkden (2015). For a more sympathetic view, see Lightfoot 2016,[3] where I claim that the analysis is being rejected and discarded prematurely.

New I-languages emerge when the ambient E-language experienced by children changes. E&F offer an intriguing sociopolitical history of language used in medieval England, addressing several matters that have been raised by traditional historians of English. They discuss how England was subjugated by Scandinavians for two hundred years and then both the English and Scandinavians were oppressed mercilessly by the Normans. By 1100, all property of any note was in the hands of the Normans and "two previously separate peoples became united in servitude" (p. 41). "The miserable circumstances gave rise to a complete fusion of two previously separate populations, speakers of Old English and speakers of Scandinavian" (p. 43). They intermarried and there was much bilingualism (O'Neil 1978). This, and not when the Scandinavians first arrived and constituted the ruling class, is when we begin to observe significant Scandinavian influence on the native written language. Despite both Germanic populations being dispossessed, the Scandinavians predominated in trade, in agriculture, and in leading the opposition to the French. The Scandinavians settled permanently in the East Midlands and North and seem to have enjoyed notably higher economic status than the native English. A plausible sociopolitical history of language in England enables us to understand better how eventually the dominant features of external language, both spoken and written, came to be Norse for many people. It explains why children came to acquire Norse syntax, given that their ambient E-language had changed and now incorporated many aspects of Norse syntax. And if they acquired elements of Norse syntax, we understand why there was a wholesale introduction of new constructions in the written language, as Old English became more and more restricted to impoverished and illiterate peasants and eventually died out in a kind of "language replacement," discussed in Campbell 2015.

The external history that E&F provide enables us to understand how the new I-languages that they postulate might have been discovered and selected by children, but the substance of their case lies in the linguistic analyses they outline. They tease apart syntactic characteristics of West Germanic (Dutch, Frisian, High and Low German, and their later offshoots Yiddish and Afrikaans) and North Germanic (mainland Scandinavian: Norse, Norwegian, Danish, Swedish), all well-analyzed languages, and they argue that Old English, to a large extent, has the characteristics of the former but Middle English the latter. For example, North Germanic has underlying head-initial VPs, West Germanic "at least partly" head-final VPs; North Germanic has infinitives with a *to* free morpheme, while West Germanic uses inflection; North Germanic has subject raising but West Germanic does not; restrictive relatives are introduced by invariant morphemes in North Germanic but by pronouns marked for case in West Germanic. E&F examine twenty such structures, which they call "parameters," and argue that early Middle English speakers began to set them in the North Germanic fashion. For all twenty parameters, they offer a plausible demonstration that Old English speakers set them in the West Germanic fashion and early Middle English speakers in the North Germanic fashion. Furthermore, Holmberg 2016 notes strikingly about E&F that "as for syntactic features that Middle English shares with Old English that are not shared with Norse, they don't find any!" Holmberg also notes that there seem to be other ways in which Middle English adopted Scandinavian characteristics, beyond what E&F consider, for example the deletion of complementizers and the avoidance of *that*–trace violations.

Whatever the details of each of the twenty structures, there was clearly a major discontinuity between Old and Middle English. The commentators are clearly strongly committed to the idea that languages change only slowly. But perhaps more will emerge in future discussions. For example McWhorter 2016 suggests that E&F gloss over factors that indicate that Middle English is too different from Norse for their hypothesis to be sustained.

There is much to be said about all twenty features, but it is noteworthy how little is said about them in the commentaries offered so far and how things that are said often turn out to be misconceived—see below. Peter Trudgill (2016) is persuaded that "Emonds and Faarlund have brilliantly demonstrated that the syntax of my native language owes a great deal to the syntax of [Old Norse]—and very much more than has generally been thought." Noting E&F's "deeply impressive achievement," he laments "the generativist mindset" of the authors, but aspects of the generativist mindset not called on by E&F in fact strengthen their case, namely the E-language–I-language distinction and matters of language acquisition.

E&F offer excellent discussion (pp. 84–93) of the very unusual property of preposition stranding, absent in most Indo-European languages including Old English and West Germanic (except Frisian and Dutch under very special circumstances) but found in the early and modern mainland-Scandinavian languages (compare also Walkden 2017 and Thoms 2019). E&F also offer nuanced discussion of the change from head-final to head-initial VPs, recognizing work showing the new V–DP order occurring sometimes in Old English texts (e.g., Pintzuk 2002).

It is important to note that, whatever the descriptive success with the twenty properties discussed, E&F attain a remarkable level of *explanatory* adequacy by postulating that the language of Norse speakers played a major role in triggering the syntactic structures of the English speech community during the early Middle English period (the twelfth and thirteenth centuries). For example, E&F discuss the verb-second properties of Old English, which show the complicating property of verb-third when the subject is pronominal and to the left of the finite verb (Van Kemenade 1987): [*ælc yfel*] he **mæg** *don*, 'each evil, he **can** do'. Whatever this special property of Old English is, it is absent from Norse and therefore, given E&F's central hypothesis, it is predictably absent from Anglicized Norse and early Middle English. By claiming that it was Norse syntactic systems that emerged, E&F *predict* that all the twenty relevant properties should have emerged in the first Middle English texts and that they should not have emerged in piecemeal fashion: all the new phenomena involve "changes" in the direction of North Germanic parameter settings. On another hand, for Bech and Walkden (2015), for example, this consistency is either accidental or not real.

E&F write, "The Old/Middle English break very much concerns the structure of the language itself; it is very little connected with how English was used or how it was perceived" (p. 28). "When English began to be written after the [Norman] Conquest, the new characteristics were clearly in the ascendant, most strongly in the former Danelaw ... while many aspects of Old English (as well as most of its vocabulary) had disappeared or been reduced to remnant percentages, especially in the South and Southwest" (p. 29).

5.5   Discovering and Selecting Elements of I-Languages

It is common within generative circles to view children as comparing the success of grammars in generating structures that match those elicited in the ambient language. An alternative stance is to embrace the antipositivist arguments of Chomsky 1975, that there is no procedure to guarantee that scientists discover or decide on correct theories; the best they can do is to compare theories and decide which one gives a superior account of some predetermined data, is learnable, achieves greater depth of explanation, and so on. A third view is that children are not constructing theories but subconsciously acquire a system that characterizes their linguistic capacity. We have argued here that they may be following a kind of discovery procedure but are limited to having structures generated by the Project and Merge operations allowed by UG.

Before we come to the discovery procedure itself, let us go back
to the idea that the reason linguists have never been able to provide
a clear definition of a language as distinct from a dialect is because
the sentences of, say, English do not constitute a recursively enumerable set. To repeat, if one asks if *She might could do it* is a sentence of English, the answer is that it is a sentence of English in
Alabama but not in Alaska. Therefore, there is no English such that
it is something acquired by children growing up in Alabama,
Alaska, and Alice Springs. Rather, children acquire different capacities, different I-languages, and one needs something more granu-
lated to capture this. The sociological notion of English seems to
be not relevant for an account of people's biological language capacity (but we return to this matter in §5.9).

We noted in §5.2 that Wilhelm von Humboldt distinguished languages as the creations of nations from languages as the creations
of individuals: language "is not a mere external vehicle, designed
to sustain social intercourse, but an indispensable factor for the
development of human intellectual powers... While languages
are... creations of nations, they still remain personal and independent creations of individuals" ([1836] 1971: 5, 22). We have observed
in several places that Chomsky 1986 follows in this vein and distinguishes external language and internal, individual languages.
E-language refers to language out there in the world, the kind of
thing that a child might be exposed to, an amorphous mass.
I-language, on the other hand, refers to a biological system that
grows in a child's mind/brain in the first few years of life and characterizes that individual's linguistic capacity. It consists of structures, categories, morphemes, phonemes, and a set of computational
operations that copy the items, delete them, assign indices to them,
and so on. One's I-language is a private object that permits communication with certain other speakers, although it is not necessarily identical to the I-languages of those speakers.

These ideas of E-language and I-languages suffice for the purposes of accounting for language acquisition, and we do not need the conventional, sociologically defined notion of English or Estonian. With these ideas, one can view children as selecting their I-language on exposure to E-language. Rather than evaluating systems against a set of data, children can be viewed as paying no attention to what any I-language or grammar generates but instead growing an I-language by discovering and selecting its elements (Lightfoot 2006a). This is a procedure whereby children exploit the UG toolbox they have at birth and parse the E-language they hear, discover the categories and structures needed to understand what they hear, and thereby select the elements of their I-language. E-language and I-languages are quite different kinds of entities but they are related: elements of I-languages are triggered by phenomena of E-language, and that is the discovery mechanism.

UG provides children with the set of structures that they might need in order to understand and parse the E-language that they are exposed to; these are the structures that robust E-language elicits in them. Children are born to parse and, as noted in §3.1, after they know that *cat* is a noun referring to a domestic feline and *sit* is an intransitive verb, they may hear an expression *The cat sat on the mat* and recognize that it contains a DP consisting of a determiner *the* and a noun *cat* and a VP containing an inflected verb *sat* followed by a PP *on the mat*. The child makes use of the structures needed to parse what is heard, and once a structure is used it is incorporated into the emerging I-language. In this way a child accumulates the structures of their I-language, which are required to parse the ambient E-language; children select elements for their I-language, the basic structures, in piecemeal fashion.

At no stage do children calculate what their current I-language can generate. Rather, they simply accumulate the necessary structures. Furthermore, if UG makes available a thousand possible

structures for children to draw from, that raises no intractable feasibility problems comparable to those facing a child evaluating the
generative capacity of grammars with thirty possible parameter settings, checking the grammars against what has been heard. There
are no elaborate calculations. Children developing some form of
English I-language learn without apparent difficulty irregular past-
tense and plural forms for a few hundred verbs and nouns. Learning that there is a structure $_{VP}$[V + Infl PP] seems to be broadly a
similar kind of learning, although much remains to be said (for
detailed discussion, see Fodor 1998a; Dresher 1999; Lightfoot
1999).

5.6 Individual Languages versus Big Data

E&F achieve considerable descriptive and explanatory success in
their account of the emergence of "Middle English," now equipped
with scare quotes. However, their account is troublesome for people who believe that language changes only gradually and imperceptibly. In contrast, we can understand why and how the changes
E&F describe should have happened if we think in terms of
I-languages being selected by individuals. We have emphasized that
I-languages exist for people and not for languages; there is no grammar of English in any biological sense. Therefore, rather than
thinking of "language change," we need to think less grossly, in
terms of the spread of new I-languages. Humboldt, Paul, and modern work take a biological view of languages, as opposed to a social
view. At a minimum, different questions arise under each view and
the same questions take on quite different complexions; this is worth
some reflection.

The overwhelmingly most common view among historical linguists is that language change is gradual. That view seems to drive
E&F's antagonists in *Language Dynamics and Change*. However,
much depends on the units of analysis, the kind of lens used. Lan-

guages seen as social entities can be seen as changing gradually, but I-languages emerge abruptly in an individual speaker.

Fries 1940 looks through a wide-angle lens and finds that Old English around the year 1000 shows object–verb order 53 percent of the time. That order is gradually replaced by verb–object order, reducing to 2 percent by the year 1500. Fries provides one set of statistics for each century but offers no analysis. His counts ignore the distinction between matrix and embedded clauses, where word orders are different, and he has no explanation for the fact that the finite verb often appears in second position in simple clauses. If one makes such distinctions, one can count more productively and see that Old English I-languages had object–verb order underlyingly, with a system yielding subjects in first and third position, and objects "extraposed" to the right (but the widely imitated analysis of Dutch and German, moving finite verbs to a higher C position in matrix clauses, which Van Kemenade 1987 applies to Old English, is not supported). Consequently, we find object–verb order uniformly in embedded clauses, but only variably in matrix clauses. In fact, at least two distinct changes took place in I-languages at different times: object–verb order was replaced by verb–object, and the operation moving objects to the right of the verb was lost (Haeberli 2002a,b).

If units of analysis are as gross as Fries's, change will look gradual. But one must be wary of "big data," often gathered these days through digitized corpora that do not make the E-language–I-language distinction that is essential to our view. In gathering data from a certain period, say fifteenth-century England, one must resist the temptation to assume that there must be a device that will generate the entire collection of data. One must be ready to distinguish abstract structures generated by new I-languages and the raw data of E-language that might trigger those new I-languages.

At the other end of the scale, if we use a telescopic lens, the speech of no two people is identical, change is everywhere, all is in flux, languages are constantly changing in piecemeal, gradual, and minor fashion; again we see constant, gradual change. Initial experiences are never entirely the same for two speakers, and they may differ in minor and insignificant ways. A particular construction type might become more frequent, perhaps as a result of taking on some expressive function. This does not reflect the properties of an I-language itself but the way in which I-languages are used. Such changes in frequency do not reflect a change in I-languages, but they do entail a change in the external language for the next generation of speakers, therefore for the PLD triggering the next I-languages.

Not only may E-language change gradually, the very nature of language acquisition ensures a kind of gradualness in that children experience the speech of their parents, older siblings, and other household members. This gradualness works against major discontinuities in the class of expressions and their associated meanings. For example, one generally does not find an I-language that yields more or less uniform object–verb order being replaced abruptly by one that yields uniform verb–object order. Even so, one does find significant discontinuities, as E&F have shown dramatically, especially in contexts where the output of a parent's native I-language does not contribute as significantly to a child's experience as the I-languages of children who have already acquired elements of Norse syntax. To be sure, there is no reason to believe that there is any formal relationship between the I-languages of parents and children. I-languages are created afresh by every individual and may differ in form from those of their parents, perhaps radically, within the limits of UG. For discussion, see Lightfoot 1999: chap. 4, where I-language changes are viewed as Thom-style catastrophes or phase transitions; there may be gradual change in the temperature of water until there are structural changes only at the critical points where it becomes a solid or gas (Thom 1972; Casti 1994). Now see Haeberli and Ihsane (to appear) for interesting new evidence corroborating one of the best known cases of reanalysis in the history of

English, that of the modal verbs (§2.4).[4] Similarly, E-language may change gradually and trigger a new I-language at a certain point (see Westergaard 2017 for discussion of the gradualness of change from a range of theoretical perspectives).

Whether differences between I-languages are small-scale or large-scale, these differences do not have temporal properties, and changes in I-languages cannot be "gradual": a person either has a particular I-language or another one. Apparent gradualness does not reflect an I-language property but a mirage, conjured by a failure to distinguish independent change events in E-language from changes of I-language. E-language may differ in ways that do not trigger a new I-language. However, a natural way for linguists to think of catastrophic changes is to envisage E-language crossing a threshold, which triggers a different I-language system. The relatively few variable properties given by UG (the abstract structures) constitutes the set of "fixed-point attractors," familiar from chaos theory (see Kauffman 1995) defining the nature of possible changes. An I-language either has some property or does not.

5.7   Back to Discontinuities

Equipped now with these ideas of E-language and I-languages, seeing language acquisition as reflecting a drive to parse E-language and to discover and select the elements of I-languages, one can understand historical discontinuities as new I-languages triggered by new E-language, that is, by new PLD. Children parse the ambient E-language and select the structures of the emerging I-languages that coexist in a single speech community and in a single speaker. Discontinuities do not constitute a special event. Rather they are the normal state of affairs confronting any child. There is no "imperfect learning" or "imperfect transmission," just different transmission. New E-language triggers new I-languages. New E-language is the initial locus of change, and furthermore, we can link particular aspects of E-language to new I-languages. Nothing is transmitted. Elements of I-languages are selected afresh in each generation and in each individual.

If work on language change provides insight into linking particular aspects of E-language to new I-languages, that constitutes a major contribution to our understanding of children's language acquisition, not achieved so far in experimental work on children. Furthermore, as in so many other domains, observing how a thing changes often reveals properties of that thing. To see how that works, let us elaborate on two changes in the I-languages of English speakers, discussed preliminarily in chapter 2. In §2.4 we noted that by Early Modern English, but beginning earlier, I-languages had *can, could, may, might, must, shall, should, will,* and *would* categorized or parsed as Infl items, whereas they had been verbs for earlier speakers. As a result, they came to have a more restricted syntactic distribution, ceasing to occur in various contexts. There is good reason to believe that this change in I-languages was triggered by a prior change in E-language.

Old English I-languages showed many inflections, indicating the tense, person, number, and conjugation type of verbs and the number, case, and declensional class of nouns. While regular Modern English verbs show up in four forms (*refuse, refuses, refused, refusing*), Old English verbs had over a hundred forms, as observed in §2.4. However, all of that was vastly simplified over the course of Middle English. The morphological distinctions were eliminated first in the north of England (manifested first in the Lindisfarne Gospels) and later in London and the south, arguably an effect of widespread English–Scandinavian bilingualism (O'Neil 1978). Individuals with English and Scandinavian I-languages had a rich morphological system, similar to each other but different and not learnable as a single system.

The only aspects of present-tense verb morphology to survive the great Middle English simplification were the third-person singular ending -*s*/-*eþ* and the second-person singular ending -*st*. It is quite unclear why those elements of present-tense verb morphology should have survived. However, the verbs that were to be recategorized as Infl elements had been members of the preterite-present class, which had what were usually past-tense endings for forms of the present tense, a phenomenon that also occurs in Latin verbs like *coepi* 'begin', *odi* 'hate' and Greek verbs like *oida* 'know' and *eoika* 'seem', which are present tense in meaning but perfect in form. The crucial fact about the morphology of the preterite-presents is that they never had the -*s*/-*eþ* third-person singular ending and therefore now lacked what was to become the single morphological property of present-tense verbs. Of the original preterite-presents, some dropped out of the language (e.g., *unnan* 'grant', *benugan* 'suffice'), others assimilated to regular verbal inflections with the third-person -*s* ending (*witan* 'know', *dugan* 'be of value'), and the remainder were recategorized as Infl items.

Furthermore, with the loss of subjunctive endings as part of the morphological impoverishment, another defining property of verbs came to be that -*d* forms indicated past tense. However, the -*d* forms of the items to be recategorized, *could, might, must, would, should*, rarely indicated past tense (**She might lift eighty kilos until yesterday*) but rather retained "subjunctive meaning."

One of the tasks of a child developing his/her I-language is to identify the words and the categories to which they belong, and this is done on the basis of formal and distributional properties and is the role of parsing. After the simplification of complex verb morphology, having the third-person singular -*s* ending became a defining property of verbs, but *can, could, may*, and so on did not have it. Verbs had past tenses in -*d*, but *might, would, should*, and *must* almost never carried past-time meaning and *could* only rarely did (*She could lift 75 kilos until she turned thirty*). Therefore, after the great simplification of morphology, the old preterite-presents lacked what had become the defining formal properties of verbs. They could no longer be verbs and were assigned to the only plausible category that could be immediately followed by a VP: Infl or T, depending on one's theory of functional categories. Hence the newly restricted syntactic distribution. If *can*, *may*, and so on were instances of Infl or T, they could not occur to the right of an aspectual marker, a position restricted to lexical verbs (**She has could gone*; **Canning go, she left angrily*). They cannot occur with the infinitival *to* or with another modal verb, which are also instances of Infl/T (taken as a cover term for the relevant functional heads), because there can be only one such element in a clause (**I want to can leave*, **She might could read it*).

New E-language resulted from the contact between English and Scandinavian speakers in the north and the associated bilingualism, if O'Neil 1978 is right. That new E-language triggered new I-languages, with the old verbs *can*, *may*, *shall*, and so on recategorized as Infl items (as shown in Lightfoot 2017a). Hence the discontinuity and its explanation. There is no imperfect learning or imperfect transmission, just new E-language triggering a new I-language. The explanation for the change is local; it involves particular changes that took place in the ambient E-language at this time. The equivalent change has not taken place in closely related languages like Dutch and German or slightly more distant languages like French and Italian, because there were no comparable changes in E-language. Furthermore, we are not trapped into the circularity of claiming that the new I-language was triggered by exposure to the new data generated by the new I-language, a problem that confronts proponents of evaluation approaches to acquisition, as we saw in chapter 2.

Let us turn now to the second of the famous and well-understood reanalyses discussed in chapter 2 (§2.5), the loss of V-in-I structures, $_{Infl}V$, resulting from the movement of a verb to an Infl position. This loss is another change that has not taken place in closely related languages like Dutch and German. Under the discovery-

and-selection approach to acquisition, for such a structure to be discovered by children, they must *need* it in order to understand and parse utterances. For example, once Chaucer's son knew that *understand* was a verb and could occur in high positions in expressions like *Understands she this chapter?* or *She understands not this chapter*, he would know that the verb could occur in an Infl position and therefore also, for the first of these two utterances, in a C position. Such sentences in E-language expressed the $_{Infl}V$ structure. Sentences like the first also expressed the $_C[_{Infl}V]$ structure. Similarly, Thomas More, for whom *can* and *may* and so on were verbs, would also understand *She cannot understand this chapter* as expressing the $_{Infl}V$ structure, because the verb *can* occurs to left of the negative *not* and therefore must have moved to the higher Infl position. Likewise, a French-speaking child hears *Elle comprend pas ce chapitre* 'She understands not this chapter' as expressing the $_{Infl}V$ structure. As does a Dutch child hearing *Begrijpt zij dit hoofdstuk?* 'Understands she this chapter?' *Comprend* has moved to the left of the negative marker, to Infl, and *begrijpt* has moved to an Infl position, from which it can move to a higher C.

However, somebody younger than Thomas More, for whom *can*, *may*, and so on were not verbs but Infl items, would not parse *She cannot understand this chapter* as expressing the $_{Infl}V$ structure, because for them *can* would be an Infl item and therefore not a verb in the Infl position. Given that a clear majority of simple sentences contain a modal auxiliary (Leech 2003), this means that there was a large reduction in the expression of the $_{Infl}V$ structure.

Alongside the fact that sentences with a modal auxiliary no longer express the $_{Infl}V$ structure, expressions like *Understands she this chapter?* and *She understands not this chapter*, which reflected the $_{Infl}V$ structure, were giving way to forms with the periphrastic *do*: *Does she understand this chapter?* and *She doesn't understand this chapter*, where the $_{Infl}V$ structure is not expressed (*do* is not a verb moved to Infl; in these examples *understand* is the verb). These forms entered the language in the late fifteenth century and spread from the southwest across the rest of the country. Ellegård 1954 provides a remarkably detailed account of that spread, and McWhorter 2009 argues that periphrastic *do* arose in the southwest under the influence of Cornish. ("Cornish's auxiliary *do* presents a thoroughly plausible model for English's periphrastic *do*" as a carrier of tense in negated and interrogative contexts [p. 164]. McWhorter goes on to show how this provides a good explanation for the spread of periphrastic *do* in Middle English.) The spread of periphrastic *do* further reduced the expression of the $_{Infl}V$ structure.

The evidence suggests that new E-language, stemming from the combination of the recategorized Infl items (an I-language change) and the new *do* forms, reduced the expression of the $_{Infl}V$ structure to below the threshold that would enable children to add it to the structures making up their mature I-languages. As a result of that change in I-languages, there were further changes in E-language, such that expressions like *Understands she this chapter?*, *She understands not this chapter*, and *She understood on Tuesday the chapter* no longer occurred.

In these two well-understood phase transitions, we see discontinuities that can be explained as responses to new E-language. In the case of the new Infl items, new E-language arose because of a prior change in I-languages, the dramatic simplification of morphological endings resulting from English–Scandinavian bilingualism. In the case of the loss of the operation moving verbs to a higher Infl position, E-language changed as an effect of the new Infl items just mentioned and of contact with Cornish, such that the old $_{Infl}V$ structure was no longer expressed sufficiently to be attained by children at this time. Nothing comparable has happened in the E-language that Dutch or German speakers experience, and therefore no comparable change in I-languages has occurred. In the case of the loss of the $_{Infl}V$ structure, however, unlike in the case of the recategorization of certain verbs, there are comparative data from other languages. Heycock et al. 2012 reports the tail end of the loss of $_{Infl}V$ structures in Faroese, and Vikner 1995 argues that mainland Scandinavian has lost those structures.

We see changes in I-languages, discontinuities, that can be understood as responses to new E-language. There is nothing imperfect at work. Of course, no two children have exactly the same PLD; they hear different things. Nonetheless, despite variation in experience, children often attain the same mature I-language in terms of the set of known syntactic structures. Individual experiences may vary indefinitely, but I-languages show structural stability and vary only in limited ways. I-languages emerge in the usual manner on exposure to E-language that has changed in a critical way, so that the discontinuity occurs naturally enough. Neither E-language nor I-languages get transmitted, neither imperfectly nor perfectly. In particular, I-languages are not transmitted in any sense. Rather, they are invented afresh by young children on exposure to the PLD in E-language. They are not shaped by earlier I-languages except indirectly through E-language. As we noted, I-languages are not objects floating smoothly through time and space.

5.8   Gradual Change in Languages or Spread of I-Languages

We have emphasized that grammars, I-languages, exist for people and not for languages; there is no grammar of English in any biological sense. We noted antecedents for this view in nineteenth-century writings. Like Humboldt and Paul, we take a biological view of languages as opposed to a social view. At a minimum, different questions arise under the two views and the same questions take on quite different complexions; for discussion, see Lightfoot 1995, 1999: 79–82.

Everyday common sense suggests that, if there is little distortion, the patterns, processes, and structures of life do not change very much. Similarly with language. In addition, the overwhelmingly most common view among historical linguists is that language change is gradual (cf. Westergaard 2017). But things depend on the units of analysis: languages seen as social entities change gradually, but I-languages change abruptly.

PLD may change gradually at first, and the very nature of language acquisition works against major discontinuities in the class of expressions and their associated meanings. For example, one does not find an I-language that yields more or less uniform object–verb order being replaced abruptly by one that yields uniform verb–object order. Even so, one does find significant discontinuities, especially in contexts where the output of a parent's native I-language does not contribute significantly to a child's PLD. To be sure, there is no reason to believe that there is any formal relationship between the I-languages of parents and children. I-languages are selected anew by every individual and may differ in form from that of their parents, perhaps radically, within the limits of UG. Whether differences between I-languages are small-scale or large-scale, they do not have temporal properties and cannot be gradual. Apparent gradualness of change is a mirage, conjured by a failure to distinguish independent change events, changes in E-language and changes in I-languages. PLD may differ in ways that do not trigger a new I-language. However, a natural way for linguists to think of significant changes is to consider different sets of PLD as sometimes crossing thresholds, which trigger different I-language systems. So the inventory of variable properties given by UG (the variable structures that may or may not occur in I-languages) constitutes the set of "fixed-point attractors," that is, the set of I-languages, defining the set of possible changes.

The ideas of Kroch and his associates on competing grammars (Kroch 1989) factor into thinking about apparent gradualness of change and the diffusion of new I-languages. Paul thought that there were as many languages as individuals (1880: 31), but Kroch and colleagues (Kroch 1994; Kroch & Taylor 1997) maintain that there are even more languages than individuals, because people operate with coexisting I-languages in a kind of internalized diglossia, indeed an internalized multiglossia. Their work enriches grammatical analyses by seeking to describe the variation we find within texts and across different writers, the spread of a grammatical change through a population. In postulating two or more coexisting I-languages in an individual, a researcher needs to show not only that the I-languages together account for the range of expressions used by that individual but also that the I-languages are all learnable under plausible assumptions about children's PLD. Multiglossic grammars are subject to exactly the same learnability demands as any other biological grammar.

This kind of diglossia provides an interesting approach to solving significant learnability problems. It offers a way to eliminate the unlearnable distinction between optional and obligatory operations (Lightfoot 1999: 92–101.). Chomsky 1995 argues that grammars do not permit optional operations. In that case, apparent optionality would be a function of coexisting I-languages. Instead of one I-language generating forms *a* and *b* optionally, one would argue that a person has access to multiple I-languages, one of which generates form *a*, another form *b*. The speaker has the option at any given time of using either of the I-languages. This move reduces the class of available grammars, eliminating those with optional operations (see also Biberauer & Richards 2006; Wallenberg 2013).

A consequence of this view is that Old English texts that only sometimes show verb-second phenomena cannot be explained by endowing Old English I-languages with a device for generating verb-second order optionally. Rather, there must have been competing I-languages, one generating verb-second order and the other not. Certain speakers have access to just one of these I-languages, other speakers have access to the other I-language, and others have access to both systems in an internalized diglossia. This turns out to be a productive analysis (Kroch & Santorini 2013).

On the view developed by Kroch, "change proceeds via competition between grammatically incompatible options which substitute for one another in usage" (1994: 180). One reason for believing that this view is along the right lines is that alternating forms cluster in their distribution; the clustering follows from how sets of I-languages unify the forms. We find not free variation but oscillation between two (or more) fixed points. This is reflected in the Constant-Rate Effect of Kroch 1989 and was noted in §2.5 in the context of Shakespeare using the old $_{Infl}V$ system as well as the new system without verb movement, alternating between the old and new systems within the same sentence.

Because structures are abstract, changing one element of structure or one categorization may entail a range of new surface phenomena. The Constant-Rate Effect requires that all surface phenomena reflecting the new I-language property have usage frequencies that change at the same rate, though not necessarily at the same time. This is easy to understand if one I-language is replaced over time by another, and if that change takes place in a winner-take-all competition between the two systems. We do not find complex arrays of linguistic data changing randomly. Instead, they tend to converge toward a relatively small number of patterns or attractors, in a kind of "antichaos" in the sense of Kauffman 1995. The points of variation defined by UG constitute the attractors and the two competing I-languages define the points of oscillation. Changes may occur when one I-language replaces another or when a new I-language is selected to coexist with another.

When we view an individual's language capacity as characterized by one or more private, personal I-languages, then the spread of a new I-language across a speech community can be approached through the methods of population biology. An individual may be exposed to PLD that differ from what anybody else has been exposed to. This could happen because of population movements, new patterns of bilingualism, or adult innovations by parents, caregivers, or teachers, or perhaps because the PLD are truncated in some way, not including earlier expressions or not including them with the same frequency as a generation earlier. If one individual selects a structure differently from others in the community, she is likely to produce different utterances. These new expressions, in turn, affect the linguistic environment, and our innovator will now be an agent of further change, reinforcing the PLD that might trigger another instance of her new I-language in younger siblings. As the younger siblings pick the same structures as their elder, so other people's PLD will differ, and a chain reaction is created. In this way a new I-language may spread in a way analogous to what has been observed in population genetics, replicating aspects of evolutionary change.

5.9 Population Dynamics and a Computational Model

Niyogi and Berwick 1995, 1997 present a computational model that analyzes change in this way and derives the trajectory of changes. Niyogi 2006 enriches the model. The model is based on a learning theory with three subcomponents: a theory of grammar, a learning algorithm by which a child generates grammars on exposure to data, and PLD. Niyogi and Berwick assume a population of child learners, a small number of whom fail to converge on preexisting grammars. After exposure to a finite amount of data, some of these children now converge on the preexisting grammar, but others attain a different I-language.

The next generation will therefore no longer be linguistically homogeneous. The third generation of children will hear sentences produced by the second—a different distribution—and they, in turn, will attain a different set of grammars. Over successive generations, the linguistic composition evolves as a dynamical system. (Niyogi & Berwick 1997: 2)

Emergence of a new I-language, in this simulation, is a logical consequence of specific assumptions about the theory of grammar, the learning algorithm, and the PLD. Interestingly, Niyogi and Berwick's model yields different time courses for different changes. A common trajectory is the familiar S-curve (Weinreich, Labov, & Herzog 1968; Kroch 1989): a change may begin gradually, pick up momentum, and proceed more rapidly, tailing off slowly before reaching completion. The success of Niyogi & Berwick is to build a dynamical system from a parameterized system and a memoryless learning algorithm. As a result, they *derive* the S-curve rather than building it into their model as a specific assumption. Further, the model allows that changing elements of the theory of grammar or of the learning algorithm may yield different trajectories, including trajectories other than an S-curve. Thus the model may be amended depending on how it matches the observed trajectory for specific changes in specific languages. This offers a new empirical demand for theories to meet, in addition to demands of learnability, coverage of data, explanatory depth, and so on: theories can be expected to provide the most accurate diachronic trajectories for identified changes.

Niyogi and Berwick 2009 recognizes that Chomsky 1965's idealization of speaker-hearers who use a homogeneous, social language led to a standard model of acquisition under which one PLD source developed into a single target grammar. Changing that idealization elicited population-based acquisition models. Niyogi & Berwick's model accurately captures how new I-languages progress through a community of speakers. This remarkable result clearly could not be replicated under a social definition of grammars, which denies the usefulness of individual, biological grammars. There may be slowness and gradualness in the spread of a change through a population, but changes in I-languages are rapid and abrupt at the individual level; familiar S-curves generally arise as a function of averaging across groups. The rapidity of change follows from an extension of the intuition behind Mark Aronoff's Blocking Constraint, which limits coexisting forms to those that are functionally distinct (Aronoff 1976).

This all strongly suggests that structural changes are rapid and abrupt at the individual level and often spread through a population rapidly. The speed of the spread depends on nongrammatical factors relating to social cohesion, facility of communication among different groups, and so on.

5.10 Conclusion

The overwhelming consensus among historical linguists is that languages change gradually. It is granted that there are significant discontinuities that happen from time to time, but they are allegedly due to unusual events. We have outlined a different approach here in terms of a distinction between amorphous and constantly shifting E-language, on the one hand (constantly in flux and never experienced the same way by any two people), and biological I-languages on the other hand (represented in the minds/brains of individuals, recursive systems that characterize the individual's language capacity). We view an I-language as emerging in a child as its elements are expressed in the ambient E-language and are selected by the child. I-languages are internal, individual entities, and speakers typically operate with more than one I-language. Different I-languages may be selected when children are exposed to different E-language.

Construing a person's language capacity as an individual, private matter, we can understand how different linguistic experience may trigger a different internal system, which may then spread through a speech community in ways that can be modeled through the methods of population biology. Indeed, the spread of many things, from political views to fashion to support of particular soccer teams, has been modeled in Niyogi's work. Under this view, discontinuities, new I-languages, are liable to emerge at any time and can be understood as natural phenomena. Nothing is transmitted from one generation to another. Children develop an I-language when exposed to E-language that expresses certain structures. A child might develop a novel I-language, which may spread through a community. That is our understanding of "language change," a derivative function that is best understood as an individual phenomenon that may affect the linguistic experience of others and lead to a shift in group behavior. By seeking to understand the emergence of new linguistic patterns through the acquisition of language systems by individuals, we can sometimes explain the new group behavior.

In this chapter we have emphasized the development of language systems as a property of individuals, sharing Hermann Paul's view that languages belong to individuals. However, a paper by Hariharan Narayanan and Partha Niyogi (2013), written just before Niyogi's untimely death, takes an intriguing approach, seeking to model how a group of linguistic agents might arrive at a shared communication system through local patterns of interaction, for example developing a shared vocabulary. If successful, this would derive the distinct group properties of, say, Navajo, Norwegian, and Nubian and thereby give them a basis in biological reality.

Understanding the emergence of new I-languages as a function of language acquisition leads us to different understanding of the apparent gradualness of change, recognizing discontinuities as a natural part of language history. I-languages are selected afresh by every individual on exposure to the ambient E-language, which is different for every individual. Languages do not get transmitted and language change is an epiphenomenon of individuals selecting their private I-languages.

The study of diachronic syntax is in its infancy. As it matures, through integration with the study of language acquisition, one would expect it to cast new light on lay ideas about languages and on their sociological character. And much more.

6 Variable Properties in Language: Their Nature and Selection

6.1 UG Is Open

So children are born to parse and they do so initially with the toolkit provided by a simple, parsimonious UG. For example, by virtue of that toolkit, children build binary-branching structures using the bottom-up Project and Merge operations. Heads project to phrasal categories, and Merge combines two elements into one element with two parts. A DP may carry an index to help indicate its referential possibilities, and indexical relations must conform to the demands of the syntax–semantics interface, particularly the principles of the Binding Theory (§4.1). Categories may be copied and deleted, subject to certain limits, and deletion is subject to the demands of the syntax–sound interface, particularly the principle that elements with no phonetic content (deleted items) be understood in a structurally prominent position, namely the complement of an overt, adjacent head (§4.2). Work by Minimalists over the last few decades has greatly simplified and "naturalized" our ideas about UG.

These tools enable children to begin to assign abstract linguistic structures to what they hear, to parse their ambient E-language, and this, in turn, determines many aspects of the meaning of expressions. Parsing is central and is carried out using the toolkit provided initially by UG and then also by the emerging I-language. The developing I-language, incorporating the invariant principles of UG, is the parsing mechanism; there is no separate cognitive device specialized for parsing, as if parsing was a separate part of the language faculty. Invariant properties stem mostly from UG and from parsing at the two interfaces. Meanwhile variable properties stem largely from parsing of E-language. Invariant properties of UG coexist with the variable properties of different I-languages. This is how I-languages are selected and elaborated, in turn enriching the parsing functions. If a child hears heads preceding their complements, she selects a property for her I-language such that the language is head-initial; other people's I-languages may be head-final. If she hears words that have the distribution of Infl elements, like English *can* and *must*, she categorizes or parses those words as instances of Infl, unlike their French translations *pouvoir* and *devoir*, which are verbs. If *can* and *must* are instances of Infl, then they have distinctive syntactic properties that distinguish them from verbs and other heads. The phenomena of the child's E-language trigger elements of the emerging I-language and induce the child to gradually enrich that I-language, selecting the elements of her internal, individual, private system.

If her younger brother experiences different E-language, perhaps because the family has moved to a different city, he may select a somewhat different I-language, because UG is open, open to the demands of her E-language and to those of her brother's different E-language.

UG provides the tools that enable the initial parsing, and then the demands of the emerging I-language begin to have a shaping effect. However, under the view adopted here, UG does not provide a set of predefined binary structural parameters that the child "sets" by evaluating and comparing the generative capacity of various candidate I-languages, selecting the most successful. Rather, our UG is open, consistent with a very wide range of variable properties that syntacticians have observed across many languages. UG, indeed, enables children to parse the variable properties of their E-language as they select their personal I-language.

Children select skeletal structures that are elaborated and fleshed out with the variable properties that result from parsing. They do not calculate what grammars can generate, nor do they evaluate grammars or set UG-defined parameters. Treating children as language-acquisition devices that evaluate and rank the "success" of multiple I-languages in generating a target language has not proven successful. If there are thirty or forty independent parameters, there are a billion or a trillion possible I-languages, each generating an infinite number of expressions. The memory capacity that such an approach requires is astronomical and well beyond the limits of imaginable feasibility.

That parametric vision discussed in §1.2, where children are viewed as setting UG-defined parameters, incorporates a vision of the variation between I-languages being highly restricted, limited to binary options. But that is not at all what we find. Instead, we find new I-languages emerging that fall into a wide range of types, with phenomena often not found in neighboring or closely related languages. The variation found can be understood as resulting from children parsing highly variable E-language. We have seen several examples of very idiosyncratic properties emerging in the history of English, for example in §2.4–§2.6, where we can understand how the new I-languages might have emerged through the parsing that seems to have been involved.

We have shown how children discover very specific things through parsing their E-language:

We have seen how *may, could* and some other verbs came to be parsed as Infl elements in Early Modern English, no longer categorized as verbs, as they were by Sir Thomas More: §2.4.

We have observed that Dutch finite verbs like *begrijpt* and French finite verbs like *comprend* are copied into high, complementizer positions and manifest verb-second properties in Dutch and French I-languages, while their translations like *understand* in English systems are not so copied: §2.5.

We have cast light on how, in nineteenth-century English, individual forms of the verb *be* like *been* or *being* begin to show up in idiosyncratic positions, not occupied by other forms of the verb: §2.6.

We have shown how in English a few centuries earlier some forty psych verbs including *like*, *ail*, and so on came to be parsed quite differently, with a different meaning, syntax, and morphology: §3.3.

We have learned how modern Chinese-speaking children discover that *ba* is a light verb, whereas in ancient times individuals parsed such forms as serial verbs: §3.4.

We have seen that null-subject languages may have different properties, being understood by new speakers differently from in earlier generations in ways that would not be expected if language variation were a function of on–off switches on structures defined at UG: §3.5.

We can now understand how such complexities arise over the course of time and how I-languages may change from generation to generation by virtue of children experiencing different external language, which require them to parse their experience differently from in earlier generations. We have shown how these developments might have taken place, given a certain view of how children parse what they hear.

In addition, it is worth noting that viewing variable properties as reflecting binary structural parameters provided by the genetic material is not one adopted by biologists examining other kinds of biological variation nor by cognitive scientists studying aspects of human cognition other than language. No doubt this is for reasons akin to the lack of success of the Principles and Parameters approach to linguistic variation. A central point of this book has been to argue for a different vision of linguistic variation. Now let us take a look at similarities between the discovery-and-selection approach to language variation and a view taken by some evolutionary biologists. The upshot will be that linguists do not need to treat variable properties as requiring mechanisms that are quite different from what biologists postulate in other domains.

6.2   Darwin's Finches

Innovative work by evolutionary biologists has approached variation as resulting from a process similar to what this book has been arguing for. Darwin lamented ([1859] 1991) that neither he nor anybody else had ever seen a new species emerging, and he regarded that as a major failure of his theory, but the finches of the Galápagos Islands that came to bear his name have been recognized in recent decades as illustrating natural selection in progress (Weiner 1995).

Taxonomists are usually either splitters or lumpers. "Faced with the diversity of Darwin's finches, some splitters recognized dozens and dozens of species and subspecies. Some lumpers went so far as to call them all a single species" (Weiner 1994: 41). "There were so many freaks, so many misfits that broke the serried ranks in the museum drawers ... Naturalists read and reread the reports of those who had seen Darwin's finches alive, they sorted and resorted the stiff little rows of specimens in the museums and they wondered what on earth was happening on Darwin's islands" (Weiner 1994: 42).

Rosemary and Peter Grant did the early revelatory work, leading to the discovery of thirteen species of finch in the Galápagos archipelago, each with a distinctively shaped beak (Grant & Grant 1989). The Grants discovered why the different species had distinctive beaks. The particular beak of each species enabled it to eat the food that was available on its own particular island: big seeds, little seeds, tree bark, even blood (the vampire finch pecks other birds' wings and tails, wounding them and sipping their blood), depending on the island. Over time, natural selection resulted in different beak shapes that were efficiently specialized for these different types of food.

The Grants collected data that show that natural selection occurs and can be seen to occur from year to year. Indeed, they were able to implement a model that predicted which beaks should evolve on different islands, given the seeds there. For each island, the model "predicted correctly the divergent paths of evolution for the beaks of finches for every one" (Weiner 1994).

In short, finches on the Galápagos Islands typically have one of the thirteen beaks the Grants identified, and the specific beak shape is the one suitable for picking up the seeds of the island they inhabit. This specialization developed over time: initially the finch's genetic material was neutral or "open" with respect to beak size and shape, but natural selection led to further specifications such that the Grants' correlation between beak characteristics and feeding patterns emerged, reflecting new genetic information.

I suggest that the kind of variation we have seen in the syntax of different languages and in different historical stages of languages is typical of the kind of variation that inspired Darwin and the Grants. It is not the kind of variation that is subject to genetically defined limitations characterized by syntactic parameters. Rather, it reflects the openness of genetic information, the way in which the environment might enhance genetic properties.

Of course, the enhancements that we see in Darwin's finches are different from those that we see in the language of three-year-old children: the finch species have selected particular beak shapes, and that selection is inherited by their offspring, whereas the three-year-old child selecting the I-language of some form of English has selected new I-language elements, and each child has to discover their I-language anew. There is no comparable inherited change.

However, the nature of the variable properties, beak shapes and I-language elements, shows similarities, and both constitute responses to environmental factors, for instance, the availability of seeds on the home island or the available distribution of words like *can* and *must*, which leads them to be categorized or parsed as Infl elements. As with specific beak shapes on different islands, so with specific I-language elements emerging in, being selected by, three-year-old children.

So variable properties across the I-languages of the world may be seen as similar in nature to the variable properties that we see elsewhere in the biological world. And in all these cases, external factors have internal effects, whether on genetic makeup or on emerging I-languages. Variation familiar to biologists is not fundamentally different from what comparative linguists observe. Seeing the similarities may enhance communication between linguists and evolutionary biologists (and others in the wider scientific community) and between different kinds of linguists who have become used to working in their isolating silos.

We view UG as open, with its effects complemented by the very specific effects of parsing. This is analogous to biologists seeing the genetics underlying variation in beak shapes as open enough to be enhanced by the effects of natural selection. This takes us into the world of complex adaptive systems, self-organization, and variation stemming from apparently minor fluctuations and varying initial conditions in evolutionary and cell biology, statistical biophysics, and other factors (Casti 1994; Haken 1984; Kauffman 1995; Prigogine & Stengers 1997). Modeling fluctuations and noise in linguistic experiences and in the availability of seeds on the Galápagos Islands stimulates an interest, widely shared these days, in minor oscillations that yield change across many domains. Comparative linguists interested in the acquisition of variable systems by young children could bring much to those broad discussions.

We have seen how several very specific variable properties may emerge in some individuals as they parse the ambient language to which they are exposed and assign linguistic structures that are different from those that earlier language users assigned. We also noted at a number of points that observing how linguistic and other structures change often reveals something about their nature. The result is a very different vision of variable properties. Proponents of binary, UG-defined parameters (§1.2) expect to see variable properties falling into narrow classes of recurring variation. The alternative vision that we have sketched, being based on the parsing of E-language, leads us to expect greater variation in I-languages, and indeed that is what we find in examining how I-languages may change from user to user across generations: languages develop idiosyncratic properties not shared by languages that are closely related historically. UG keeps languages similar to each other in conforming to invariant properties that are part of our biological endowment. But UG is open, open enough to allow languages to vary as parsing requirements demand, when children discover new contrasts and select new I-language structures accordingly. Evolutionary biologists have found that same kind of variation in the beaks of Darwin's finches, and we expect that the parsing-based analysis we have developed and the approach to learning it entails will lead to a better understanding of language variation than the Principles and Parameters vision has yielded, one where information provided by UG is supplemented by information that emerges through learning through parsing.

Notes

Chapter 1

1. We return to the distribution of VP ellipsis in §4.2, where we will see that the restrictions on deletion in abstract structures captures a wide range of apparently unrelated phenomena.

2. Footnotes omitted.

3. Footnote omitted. Musso et al.'s results were anticipated in Smith and Tsimpli's 1995 investigations into a linguistic savant's learning of artificial languages, which differed greatly depending on whether the language conformed or not to the demands of UG.

4. The nonoccurrence of *I wonder whether has Kim visited Washington*, where subject inversion causes the structure to crash, is explained by the fact that two words, the "complementizer" *whether* and the preposed *has*, need to be in the same position, namely in the C that heads the clausal phrase, CP, that dominates *Kim visited Washington*. But there is room for only one of them.

5. Jim Higginbotham was the source of this metaphor, at the 1979 workshop in Pisa where Chomsky first presented his new thinking about government and binding (Chomsky 1981a).

6. For discussion, see Chomsky 1995: 212, n. 4.

7. The thinking seems to be that, if one takes biolinguistics seriously, one must use the current primitives of biologists, and not invoke features, *even if features are needed for linguistic description*. This prejudice

recalls debates between Chomsky and Piaget, when Piagetians argued that postulated UG constraints simply cannot be innate if specific to language, regardless of how useful they are analytically, unless they are stated in terms familiar to biologists (Piattelli-Palmarini 1980).

8. It is worth noting that the metric is technically and conceptually flawed insofar as it is based on an assumption that grammars with a greater number of parameters set correctly will be the fittest, the most successful in parsing/generating incoming data. Dresher 1999: 54–58 demonstrates that this assumption is false, that there is no smooth correlation between accuracy in what is parsed/generated and the number of parameters set correctly. See Lightfoot 2006a: 74 for discussion.

9. For good, wide-ranging discussion of parameters, see Karimi & Piattelli-Palmarini 2017. In that volume, some papers argue for dispensing with formal parameters at the level of UG, but Cinque 2017 and Rizzi 2017 offer responses to some of the criticisms. Epstein, Obata, and Seeley 2017 has proposals that overlap with Lightfoot 2017b in advocating an open UG. From the same volume, Longobardi & Guardiano 2017 is motivated by similar concerns to ours and seeks to replace the Principles and Parameters approach with a simplified model of the language faculty, which eliminates parameters altogether from UG, replacing them with a few abstract variation schemata. Also see Baker 2001, a very insightful discussion of parameters analogized to the elements of chemistry.

10. Berwick 1985 keys language acquisition to the products of a parsing device, and Fodor 1998a tackles problems in generating multiple parses. Parsing is central also for our children in their language acquisition, but in contrast and perhaps overambitiously, we postulate no specific parsing procedures distinct from I-language elements.

11. One point worth noting here, to be explored further in chapter 5, is that by not invoking UG-defined parameters we have no way to exclude wild impossibilities: a preposition followed by an IP complement, for example. However, such impossibilities would never be triggered, given that a person's E-language is generated by sets of I-languages in other speakers of the community. If such structures could never be triggered, they do not need to be specifically excluded. People who are contributing to an individual's ambient E-language are using exactly the same mechanisms as people parsing that E-language, namely I-languages subject to

UG constraints and to the triggering effects of their experience; they are not using parsing procedures independent of I-languages.

Chapter 2

1. Many people have contributed to our current understanding of these two phase transitions. Roberts 2007 gives a good, detailed textbook account of both changes in I-languages, though viewing them as changes in parameter settings.

2. Lightfoot 2006a: 57–61 discusses the Binding Theory, a vast improvement on earlier analyses of the referential properties of DPs, but a theory that nonetheless raises (solvable) learnability issues, which are not addressed in the literature prior to Lightfoot 2006a. See §4.1 here.

3. For more discussion of how a particular structural element may be triggered by quite different PLD in different I-languages, see Lightfoot 2006a: 123–136.

Chapter 3

1. This is not a terminological issue: thinking in terms of I-languages "changing" or being "restructured" has misled linguists into postulating DIACHRONIC PROCESSES, where one I-language becomes another by some formal operation, and then asking about the nature of those processes, whether they simplify the grammars or make them more efficient or drive them to a different type or something else along those lines. This is a topic for another discourse (Lightfoot 2017a).

2. Ambient E-language typically has several sources, including multiple I-languages, as emphasized by Aboh 2015, 2017.

3. For a more up-to-date and radical approach to ideas about abstract case, see Preminger 2014.

4. Caveat lector: the important thing is to *derive* the change in meaning of psych verbs from a property that would affect language acquisition. Here we derive the change from the loss of inherent case due to the loss of morphological case. This involves general claims about the relationship between morphological and inherent case that may need revision or elaboration. See Preminger 2014.

5. For critical discussion, see Lightfoot 1979: 224.

6. Meanwhile Huang, Li, and Li 2009: §5.4.1, (53) puts forward the very similar but somewhat simpler:

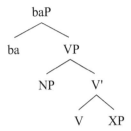

Chapter 4

1. For example, Elbourne investigates the Salience Hypothesis, the idea that the different behavior of referential and quantificational antecedents does not reflect a real difference in the way that bound and referential pronouns are analyzed with respect to Principle B; rather, it arises because children interpret pronouns as referring to the most prominent characters.

2. Elbourne 2005 and Conroy et al. 2009 dispute Thornton and Wexler's idea that Principle B violations occur only with nonquantificational antecedents. Conroy et al.'s paper constitutes a major clarification of the complex literature on the alleged "delay-of–Principle B effect."

3. Lobeck 1995 and Zagona 1988 offer good Government and Binding accounts of VP ellipsis in terms of the Empty Category Principle of more than thirty years ago; this section adopts the spirit of those analyses in requiring a host to license ellipsed VPs and, in doing so, draws on Lightfoot 2006b.

As a simple illustration of the apparent role of a host for deletion sites, Potsdam 1997 observes the distinction between (ia) and (ib,c), where *do* and *not* appear to license an ellipsed VP; Potsdam notes a similar distinction between (iia) and (iib), where *to* licenses an ellipsed VP.

(i) a. *It is possible to eat this fruit, and we recommend that you ~~$_{VP}$[eat this fruit]~~.

b. It is possible to eat this fruit, and we recommend that you do $_{VP}$[~~eat this fruit~~].

c. It is possible to eat this fruit, and we recommend that you not $_{VP}$[~~eat this fruit~~].

(ii) a. *Kim began singing a song before Jim began $_{VP}$[~~singing a song~~].

b. Kim began to sing a song before Jim began to $_{VP}$[~~sing a song~~].

4. I assume here that restrictive relative clauses are complements to nouns, distinguishing (10b, 11b); I will take nonrestrictive relatives to be noncomplements, that is, adjuncts. We need a syntactic distinction between restrictive and nonrestrictive relative clauses, and restrictive relatives have some properties of complement clauses. What I am calling complement structures may be captured through Richard Kayne's 1994 raising analysis of restrictive relatives, where a restrictive relative headed by *that* is the complement of D and the head raises out of the relative clause. For a good discussion of relative clauses and the problems they pose for modern theories of phrase structure, see Borsley 1997.

5. Bošković and Lasnik 2003, adopting ideas from Pesetsky 1991 (an unpublished extension of Pesetsky 1995), treats null complementizers as resulting from a phonological affixation operation. For Bošković and Lasnik, affixation requires adjacency, but the data of (11) show that head–complement relations are also crucially involved.

6. Notice that *Which man did Jay introduce to Ray and Jim to Tim?*, analogous to the ill-formed (14c), is well-formed. Here only one *wh-* phrase is overt and it moves across the board. One way of thinking of this is that across-the-board movement takes place on a three-dimensional structure before the two clauses are linearized; at that point *which man* is the complement of *introduce* (Williams 1978).

7. I also adopt the proposals of Nunes 2004, namely that deletion of the copied element follows from the linearization of chains. Linearization is a phonological operation that converts a syntactic structure into a sequence of items in consonance with the Linear-Correspondence Axiom of Kayne 1994. The two *what*s in a structure like (i) are nondistinct, and this leads to ordering contradictions. *What* must precede *buy*, for instance, but *what* must also follow *buy*. That is a contradiction: x cannot both precede and follow y.

(i) [what [did $_{IP}$[you ~~did~~ buy ~~what~~]]]

It is this failure to yield a linear order that renders the structure ill-formed—unless one of the *what*s is deleted; and it must be the lower *what*, for reasons of the Binding Theory. So, the fact that there can be no chains in the phonology with more than one overtly realized link entails that the lower *what* in (i) must be deleted. Nunes offers a rich analysis, noting exceptional cases where multiple *wh-* items are pronounced; there he shows that it is only intermediate copies that may be pronounced, not the lowest copy, and, indeed, that these pronounced copies have clitic-like qualities (see Nunes 2004: 38–43 for discussion).

8. There are many interesting distinctions at work. Compare, for example, *I wonder what is/'s that up there*, where reduction is possible. In this example there is no deletion site right-adjacent to *is*, and so *is* may be reduced. In (16c), however, the deletion site of *what* is between *is* and *up*, blocking reduction.

9. Another category of deletions different from ellipsed VPs is "pseudogapping," when an auxiliary is present; such constructions are as bad as the gaps in (27), for example, **Which man did Jay introduce to Ray and which woman did Jim to Tim?*, analogous to (14c), or **Jay wondered what Kay gave to Ray and what did Jim to Tim*, analogous to (14d). Pseudogapping structures are often analyzed very differently (but see Lasnik 1999: chap. 7, which treats them as VP ellipsis plus "remnant raising"), and I avoid them here. A good theory of parsing would show how children make the right selections.

10. Notice that (31a) is well-formed but (9a) is not. In (31a) the null VP is the complement of the adjacent *could've*, but in (9a) it is not the complement of *John's*. This also accounts for the following distinction.

(i) a.   Kim canceled her subscription, and I would've $_{VP}e$, too.
    b.   *Kim canceled her subscription, and I'd've $_{VP}e$, too.

A null VP following the reduced *'ve* in (ia) is the complement of *would've*, but a null VP following the reduced *'ve* in (ib) is not the complement of *I'd've*, where the reduced auxiliaries have been cliticized to the subject DP and are no longer in the position of Infl with a VP complement.

11. In (ia) the complementizer contained in the clausal complement to *thought* cliticizes to *thought* straightforwardly and may be deleted (unpronounced). In (ib) the lowest *who* cliticizes to *saw*, the intermediate *who* to *think*, and the complementizer to *think +who*, successively like the clitics of (31), again straightforwardly.

(i) a.   I thought [that/0 Ray saw Fay].
   b.   Who$_i$ did you think [~~who$_i$~~ that/0 Ray saw ~~who$_i$~~]?
   c.   *Who$_i$ did you think [~~who$_i$~~ that **who**$_i$ saw Fay]?

However, in (ic), the intermediate *who* cliticizes to *think* just like in (ib), but the lowest (boldface) *who* apparently cannot cliticize to *that*, presumably because *that* does not take complements in the usual sense (despite the name "complementizer," the following clause does not "complete" the meaning of *that* in the way that *Fay* completes the meaning of *saw*) and is not an appropriate host. Likewise for equivalent complementizers in other languages.

Similarly, in (iia) the boldface *who* may not cliticize to the complementizer *how*, because *how* has no complement. Hence the difference with (iib), where each copied element is deleted in the appropriate way, and the sentence is grammatical if not completely felicitous.

(ii) a.   *Who$_i$ do you wonder [~~who$_i$~~ how [**who**$_i$ solved the problem]]?
   b.   What$_i$ do you wonder [~~what$_i$~~ how [John solved ~~what$_i$~~]]?

There is also an interesting range of comparative data resulting from the discomfort in subject DPs with respect to displacement in several languages; see Rizzi 1990: §2.6 and Lightfoot 2006b: §2.5–§2.6 on French, West Flemish, Swedish, and Vata.

Chapter 5

1. This chapter draws on Lightfoot 2017d and my review of Emonds and Faarlund 2014, Lightfoot 2016.

2. The French, of whom Thom is one, see catastrophes frequently, as noted by others; three-syllable French *catastrophes*, /kata**strof**/, with stress on the final syllable, seem less catastrophic than four-syllable English /ka**ta**strofiz/ with antepenult syllable stressed. So French and English "catastrophes" have somewhat different meanings as well as different pronunciations.

3. Much of the commentary concerns the genealogy of Middle and Modern English, but if languages are not transmitted in the way presupposed by the cladograms of people defining language "families," more radical rethinking is needed.

4. Haeberli and Ihsane consider data concerning the distribution of adverbs that has never previously been linked to properties of modal verbs.

References

Aboh, E. O. (2015). *The emergence of hybrid grammars: Language contact and change.* Cambridge: Cambridge University Press.

Aboh, E. O. (Ed.). (2017). Complexity in human languages: A multifaceted approach. Special issue, *Language Sciences, 60.*

Adams, M. (1987). From Old French to the theory of pro-drop. *Natural Language and Linguistic Theory, 5,* 1–32.

Allen, C. (1986). Reconsidering the history of *like. Journal of Linguistics, 22*(2), 375–409.

Allen, C. (1995). *Case marking and reanalysis: Grammatical relations from Old to Early Modern English.* Oxford: Oxford University Press.

Anderson, J. M. (1986). A note on Old English impersonals. *Journal of Linguistics, 22*(1), 167–77.

Aronoff, M. (1976). *Word formation in generative grammar.* Cambridge, MA: MIT Press.

Avrutin, S. (1994). *Psycholinguistic investigations in the theory of reference.* Unpublished PhD dissertation, Massachusetts Institute of Technology.

Avrutin, S. & Wexler, K. (1992). Development of Principle B in Russian: Coindexation at LF and coreference. *Language Acquisition, 2*(4), 259–306.

Baker, C. L. & McCarthy, J. J. (Eds.). (1981). *The logical problem of language acquisition.* Cambridge, MA: MIT Press.

Baker, M. C. (2001). *The atoms of language.* New York: Basic Books.

Baker, M. C. (2008). The macroparameter in a microparametric world. In T. Biberauer (Ed.), *The limits of syntactic variation* (pp. 351–373). Amsterdam: Benjamins.

Bech, K. & Walkden, G. (2015). English is (still) a West Germanic language. *Nordic Journal of Linguistics, 39*(1), 65–100.

Bejar, S. (2002). Movement, morphology and learnability. In D. W. Lightfoot (Ed.), *Syntactic effects of morphological change* (pp. 307–325). Oxford: Oxford University Press.

Belletti, A. (2017). Internal grammar and children's grammatical creativity against poor inputs. *Frontiers in Psychology: Language Sciences, 8,* 2074. https://doi.org/10.3389/fpsyg.2017.02074.

Belletti, A. & Rizzi, L. (1988). Psych-verbs and θ-theory. *Natural Language and Linguistic Theory, 6*(3), 291–352.

Berwick, R. C. (1985). *The acquisition of syntactic knowledge.* Cambridge, MA: MIT Press.

Berwick, R. C. & Chomsky, N. (2016). *Why only us: Language and evolution.* Cambridge, MA: MIT Press.

Biberauer, T. (Ed.). (2008). *The limits of syntactic variation.* Amsterdam: Benjamins.

Biberauer, T. & Richards, M. (2006). True optionality: When the grammar doesn't mind. In C. Boeckx (Ed.), *Minimalist essays* (pp. 35–67). Amsterdam: Benjamins.

Boeckx, C. (2006). *Linguistic Minimalism: Origins, concepts, methods, and aims.* Oxford: Oxford University Press.

Boeckx, C. (2014). What Principles and Parameters got wrong. In C. Picallo (Ed.), *Linguistic variation in the Minimalist framework.* Oxford: Oxford University Press. https://doi.org/10.1093/acprof:oso/9780198702894.003 .0008.

Boeckx, C. (2015). *Elementary syntactic structures: Prospects of a feature-free syntax.* Cambridge: Cambridge University Press.

Borer, H. (1984). *Parametric syntax.* Dordrecht, the Netherlands: Foris.

Borges, H. & Pires, A. (2017). The emergence of Brazilian Portuguese: Earlier evidence for the development of a partial null subject grammar. *Proceedings of the Linguistic Society of America, 2.31,* 1–15. https://doi .org/10.3765/plsa.v2i0.4096.

Borsley, R. (1997). Relative clauses and the theory of phrase structure. *Linguistic Inquiry, 28*(4), 629–647.

Bošković, Ž. & Lasnik, H. (2003). On the distribution of null complementizers. *Linguistic Inquiry, 34*(4), 527–546.

Bresnan, J. W. (1972). *Theory of complementation in English syntax.* Unpublished Ph.D. dissertation, Massachusetts Institute of Technology.

Campbell, L. (2015). Do languages and genes correlate? Some methodological issues. *Language Dynamics and Change, 5*(2), 202–226.

Carstens, V. (2002). Antisymmetry and word order in serial constructions. *Language, 78*(1), 3–50.

Casti, J. (1994). *Complexification: Explaining a paradoxical world through the science of surprise.* New York: HarperCollins.

Chao, Y. R. (1968). *A grammar of spoken Chinese.* Berkeley, CA: University of California Press.

Chomsky, N. (1957). *Syntactic structures.* The Hague: Mouton.

Chomsky, N. (1965). *Aspects of the theory of syntax.* Cambridge, MA: MIT Press.

Chomsky, N. (1975). *The logical structure of linguistic theory.* New York: Plenum.

Chomsky, N. (1980). On binding. *Linguistic Inquiry, 11*(1), 1–46.

Chomsky, N. (1981a). *Lectures on government and binding.* Dordrecht, the Netherlands: Foris.

Chomsky, N. (1981b). Principles and parameters in syntactic theory. In N. Hornstein & D. W. Lightfoot (Eds.), *Explanation in linguistics: The logical problem of language acquisition* (pp. 132–175). London: Longman.

Chomsky, N. (1986). *Knowledge of language: Its nature, origin, and use.* New York: Praeger.

Chomsky, N. (1995). *The Minimalist Program.* Cambridge, MA: MIT Press.

Chomsky, N. (2001). Derivation by phase. In M. Kenstowicz (Ed.), *Ken Hale: A life in language* (pp. 1–52). Cambridge, MA: MIT Press.

Chomsky, N. (2005). Three factors in language design. *Linguistic Inquiry, 36*(1), 1–22.

Cinque, G. (2013). Cognition, universal grammar and typological generalizations. *Lingua*, 130, 50–65.

Cinque, G. (2017). A microparametric approach to the head-initial/head-final parameter. In S. Karimi & M. Piattelli-Palmarini (Eds.), Parameters; special issue, *Linguistic Analysis*, 41(3–4), 309–366.

Clark, R. (1992). The selection of syntactic knowledge. *Language Acquisition*, 2, 83–149.

Collins, C. (1997). Argument sharing in serial verb constructions. *Linguistic Inquiry*, 28(2), 461–497.

Conroy, A., Takahashi, E., Lidz, J., & Phillips, C. (2009). Equal treatment for all antecedents: How children succeed with Principle B. *Linguistic Inquiry*, 40(3), 446–486.

Cowper, E. & Hall, D. (2019). Scope variation in contrastive hierarchies of morphosyntactic features. In D. Lightfoot & J. Havenhill (Eds.), *Variable properties in language: Their nature and acquisition* (pp. 27–41). Washington, DC: Georgetown University Press.

Crain, S. (2012). *The emergence of meaning.* Cambridge: Cambridge University Press.

Crain, S. & Thornton, R. (1998). *Investigations in Universal Grammar: A guide to experiments in the acquisition of syntax.* Cambridge, MA: MIT Press.

Darwin, C. (1991). *On the origin of species by means of natural selection.* Amherst, NY: Prometheus Books. (Original work published 1859.)

den Dikken, M. (2012). *The Cambridge handbook of generative grammar.* Cambridge: Cambridge University Press.

Denison, D. (1990). The OE impersonals revived. In S. Adamson, V. Law, N. Vincent, & S. Wright (Eds.), *Papers from the fifth International Conference on English Historical Linguistics* (pp. 111–140). Amsterdam: Benjamins.

Ding, N., Melloni, L., Zhang, H., Tian, X., & Poeppel, D. (2016). Cortical tracking of hierarchical linguistic structures in connected speech. *Nature Neuroscience*, 19, 158–164.

Dobzhansky, T. (1964). Biology, molecular and organismic. *American Zoologist*, 4, 443–452.

Dobzhansky, T. (1973). Nothing in biology makes sense except in the light of evolution. *American Biology Teacher*, 35, 125–129.

Dresher, B. E. (1999). Charting the learning path: Cues to parameter setting. *Linguistic Inquiry, 30*(1), 27–67.

Dresher, B. E. (2009). *The contrastive hierarchy in phonology.* Cambridge: Cambridge University Press.

Dresher, B. E. (2019). Contrastive feature hierarchies in phonology: Variation and universality. In D. Lightfoot & J. Havenhill (Eds.), *Variable properties in language: Their nature and acquisition* (pp. 13–25). Washington, DC: Georgetown University Press.

Duguine, M., Irurtzun, A., & Boeckx, C. (2017). Linguistic diversity and granularity: Two case-studies against parametric approaches. In S. Karimi & M. Piattelli-Palmarini (Eds.), Parameters; special issue, *Linguistic Analysis, 41*(3–4), 445–473.

Elbourne, P. (2005). On the acquisition of Principle B. *Linguistic Inquiry, 36*(3), 333–365.

Eldridge, N. & Gould, S. J. (1972). Punctuated equilibria: An alternative to phyletic gradualism. In T. J. M. Schopf (Ed.), *Models of paleobiology* (pp. 82–115). San Francisco: Freeman Cooper.

Ellegård, A. (1954). *The auxiliary do: The establishment and regulation of its use in English.* Stockholm: Almqvist & Wiksell.

Emonds, J. E. (1985). *A unified theory of syntactic categories.* Dordrecht, the Netherlands: Foris.

Emonds, J. E. & Faarlund, J. T. (2014). *English: The language of the Vikings.* Olomouc Modern Language Monographs 3. Olomouc, Czech Republic: Palacky University.

Epstein, S. D., Obata, M., & Seeley, T. D. (2017). Is linguistic variation entirely linguistic? In S. Karimi & M. Piattelli-Palmarini (Eds.), Parameters; special issue, *Linguistic Analysis, 41*(3–4), 481–516.

Fodor, J. D. (1998a). Unambiguous triggers. *Linguistic Inquiry, 29*(1), 1–36.

Fodor, J. D. (1998b). Learning to parse? *Journal of Psycholinguistic Research, 27*(2), 285–319.

Fodor, J. D. (1998c). Parsing to learn. *Journal of Psycholinguistic Research, 27*(3), 339–374.

Fries, C. (1940). On the development of the structural use of word-order in Modern English. *Language, 16*, 199–208.

Getz, H. R. (2018a). Acquiring *wanna*: Beyond Universal Grammar. *Language Acquisition, 26*(2), 119–143.

Getz, H. R. (2018b). *Sentence first, arguments after: Mechanisms of morphosyntax acquisition.* Unpublished PhD dissertation, Georgetown University.

Getz, H. R. (2019). Acquisition of morphosyntax: A pattern learning approach. In D. Lightfoot & J. Havenhill (Eds.), *Variable properties in language: Their nature and acquisition* (pp. 103–114). Washington, DC: Georgetown University Press.

Getz, H. R., Ding, N. Newport, E. L., & Poeppel, D. (2018). Cortical tracking of constituent structure in language acquisition. *Cognition*, 181, 135–140.

Getz, H. R. & Lightfoot, D. W. (To appear). Children discovering variable properties.

Gibson, E. & Wexler, K. (1994). Triggers. *Linguistic Inquiry*, *25*(3), 407–454.

Gleick, J. (1987). *Chaos: Making a new science.* New York: Viking Penguin.

Grant, R. & Grant, P. (1989). *Evolutionary dynamics of a natural population.* Princeton, NJ: Princeton University Press.

Greenberg, J. H. (1966). Some universals of grammar with particular reference to the order of meaningful elements. In J. H. Greenberg (Ed.), *Universals of language* (pp. 73–113). Cambridge, MA: MIT Press.

Grodzinsky, Y. & Reinhart, T. (1993). The innateness of binding and coreference. *Linguistic Inquiry*, *24*(1), 69–101.

Guasti, M. T. (2016). *Language acquisition: The growth of grammar* (2nd ed.). Cambridge, MA: MIT Press.

Haeberli, E. (2002a). Inflectional morphology and the loss of verb second in English. In D.W. Lightfoot (Ed.), *Syntactic effects of morphological change* (pp. 88–106). Oxford: Oxford University Press.

Haeberli, E. (2002b). Observations on the loss of verb second in the history of English. In C. J.-W. Zwart & W. Abraham (Eds.), *Studies in comparative Germanic syntax: Proceedings from the fifteenth Workshop on Comparative Germanic Syntax* (pp. 245–272). Amsterdam: Benjamins.

Haeberli, E. & Ihsane, T. (To appear). The recategorization of modals in English: Evidence from adverb placement. In B. Egedi & V. Hegedus (Eds.),

Functional heads across time: Syntactic re-analysis and change. Oxford: Oxford University Press.

Haegeman, L. & Ihsane, T. (2001). Adult null subjects in the non-pro-drop languages: Two diary dialects. *Language Acquisition, 9*, 329–346.

Haider, H. (2005). How to turn German into Icelandic—and derive the OV–VO contrast. *Journal of Comparative Germanic Syntax, 8*, 1–53.

Haken, H. (1984). *The science of structure: Synergetics*. New York: Van Nostrand Reinhold.

Heinz, J. (2016). Computational theories of learning and developmental psycholinguistics. In J. Lidz, W. Snyder, & J. Pater (Eds.), *The Oxford handbook of developmental linguistics* (pp. 633–663). Oxford: Oxford University Press.

Heinz, J. & Idsardi, W. (2011). Sentence and word complexity. *Science, 333* (6040), 295–297.

Heinz, J. & Idsardi, W. (2013). What complexity differences reveal about domains in language. *Topics in Cognitive Science, 5*(1), 111–131.

Heycock, C., Sorace, A., Hansen, Z. S., Wilson, F., & Vikner, S. (2012). Detecting the late stages of syntactic change: The loss of V-to-T in Faroese. *Language, 88*, 558–600.

Hockett, C. (1958). *A course in modern linguistics*. New York: Macmillan.

Holmberg, A. (2010). Parameters in minimalist theory: The case of Scandinavian. *Theoretical Linguistics, 36*, 1–48.

Holmberg, A. (2016). Norse against Old English: 20–0. *Language Dynamics and Change, 6*(1), 21–23.

Hornstein, N. (2009). *A theory of syntax: Minimal operations and Universal Grammar*. Cambridge: Cambridge University Press.

Hornstein, N. & Idsardi, W. (2014). The 3rd Hilbert Question: What can you learn from degree 0-ish data. *Faculty of Language* (blog). http://facultyoflanguage.blogspot.com/2014/12/the-third-hilbert-question-what-can-you.html.

Hornstein, N. & Weinberg, A. S. (1981). Case theory and preposition stranding. *Linguistic Inquiry, 12*(1), 55–91.

Huang, C.-T. J., Li, Y.-H. A., & Li, Y. (2009). *The syntax of Chinese*. Cambridge: Cambridge University Press.

Hudson Kam, C. & Newport, E. L. (2005). Regularizing unpredictable variation: The roles of adult and child learners in language formation and change. *Language Learning and Development, 1*(2), 151–195.

Humboldt, W. von. (1971). *Linguistic variability and intellectual development* (G. C. Buck & F. A. Raven, Trans.). Philadelphia: University of Pennsylvania Press. (Original work published 1836.)

Jakobson, R. (1941). *Kindersprache, und allgemeine Lautgesetze.* Uppsala: Uppsala Universitets Årsskrift.

Jespersen, O. (1909–1949). *A modern English grammar on historical principles.* London: Allen & Unwin.

Karimi, S. & Piattelli-Palmarini, M. (Eds.). (2017). Parameters. Special issue, *Linguistic Analysis, 41*(3–4).

Kauffman, S. (1995). *At home in the Universe: The search for laws of self-organisation and complexity.* Oxford: Oxford University Press.

Kayne, R. S. (1994). *The antisymmetry of syntax.* Cambridge, MA: MIT Press.

Kayne, R. S. (1996). Microparametric syntax: Some introductory remarks. In J. R. Black & V. Motapanyane (Eds.), *Microparametric syntax and dialect variation* (pp. ix–xviii). Amsterdam: Benjamins.

Kayne, R. S. (2005). Some notes on comparative syntax, with special reference to English and French. In G. Cinque & R. S. Kayne (Eds.), *The Oxford handbook of comparative syntax* (pp. 3–69). Oxford: Oxford University Press.

Keenan, E. (2002). Explaining the creation of reflexive pronouns in English. In D. Minkova & R. Stockwell (Eds.), *Studies in the history of the English language: A millennial perspective* (pp. 325–354). Berlin: Mouton de Gruyter.

Kegl, J., Senghas, A., & Coppola, M. (1998). Creation through contact: Sign language emergence and sign language change in Nicaragua. In M. DeGraff (Ed.), *Language creation and change: Creolization, diachrony, and development* (pp. 179–237). Cambridge, MA: MIT Press.

Kiparsky, P. (1968). Linguistic universals and linguistic change. In E. Bach & R. Harms (Eds.), *Universals in linguistic theory* (pp. 171–202). New York: Holt, Rinehart, and Winston.

Kroch, A. (1989). Reflexes of grammar in patterns of language change. *Language Variation and Change, 1,* 199–244.

Kroch, A. (1994). Morphosyntactic variation. In K. Beals (Ed.), *Papers from the 30th regional meeting of the Chicago Linguistic Society: Parasession on variation and linguistic theory* (pp. 180–201). Chicago: Chicago Linguistic Society.

Kroch, A. & Santorini, B. (2013, June 29). What a parsed corpus is and how to use it. Presentation, Workshop on Diachronic Syntax, LSA Summer Institute. http://www.ling.upenn.edu/~kroch/lsa13ws.html.

Kroch, A. & Taylor, A. (1997). The syntax of verb movement in Middle English: Dialect variation and language contact. In A. van Kemenade & N. Vincent (Eds.), *Parameters of morphosyntactic change* (pp. 297–325). Cambridge: Cambridge University Press.

Labov, W. (1972). *Sociolinguistic patterns*. Philadelphia: University of Pennsylvania Press.

Larson, R. (1991). Some issues in verb serialization. In C. Lefebvre (Ed.), *Serial verbs: Grammatical, comparative, and cognitive approaches* (pp. 185–210). Amsterdam: Benjamins.

Lasnik, H. (1976). Remarks on coreference. *Linguistic Analysis*, *2*(1), 1–22.

Lasnik, H. (1999). *Minimalist analysis*. Oxford: Blackwell.

Lasnik, H. (with Depiante, M. & Stepanov, A.) (2000). *"Syntactic structures" revisited: Contemporary lectures on classic transformational theory*. Cambridge, MA: MIT Press.

Leech, G. (2003). Modality on the move: The English modal auxiliaries 1961–1992. In R. Facchinetti, M. Krug, & F. R. Palmer (Eds.), *Modality in contemporary English* (pp. 223–240). Berlin: Mouton de Gruyter.

Li, C. & Thompson, S. (1974). Historical change of word order: A casestudy in Chinese and its implications. In J. M. Anderson & C. Jones (Eds.), *Historical linguistics: Proceedings of the First International Conference on Historical Linguistics* (pp. 199–217). Amsterdam: North Holland.

Lidz, J. (2010). Language learning and language universals. *Biolinguistics*, *4*, 201–217.

Lidz, J. & Gagliardi, A. (2015). How nature meets nurture: Universal Grammar and statistical learning. *Annual Review of Linguistics*, *1*, 333–353.

Lightfoot, D. W. (1979). *Principles of diachronic syntax*. Cambridge: Cambridge University Press.

Lightfoot, D. W. (1982). *The language lottery: Toward a biology of language.* Cambridge, MA: MIT Press.

Lightfoot, D. W. (1989). The child's trigger experience: Degree-0 learnability. *Behavioral and Brain Sciences* 12(2): 321–334.

Lightfoot, D. W. (1991). *How to set parameters: Arguments from language change.* Cambridge, MA: MIT Press.

Lightfoot, D. W. (1993). Why UG needs a learning theory: Triggering verb movement. In C. Jones (Ed.), *Historical linguistics: Problems and perspectives* (pp. 190–214). London: Longman.

Lightfoot, D. W. (1995). Grammars for people. *Journal of Linguistics, 31*(2), 393–99.

Lightfoot, D. W. (1999). *The development of language: Acquisition, change, and evolution.* Oxford: Blackwell.

Lightfoot, D. W. (2002). [Introduction]. In *Syntactic structures*, by N. Chomsky, 2nd ed. (pp. v–xviii). Berlin: Mouton de Gruyter.

Lightfoot, D. W. (2005). Learning from creoles. *Lingua, 115,* 197–199.

Lightfoot, D. W. (2006a). *How new languages emerge.* Cambridge: Cambridge University Press.

Lightfoot, D. W. (2006b). Minimizing government: Deletion as cliticization. *Linguistic Review, 23*(2), 97–126.

Lightfoot, D. W. (2011). Multilingualism everywhere. *Bilingualism: Language and Cognition, 14*(2), 162–164.

Lightfoot, D. W. (2012). Explaining matrix/subordinate domain discrepancies. In L. Aelbrecht, L. Haegeman, & R. Nye (Eds.), *Main clause phenomena: New horizons* (pp. 159–176). Amsterdam: Benjamins.

Lightfoot, D. W. (2013). Types of explanation in history. *Language, 89*(4), e18–e38.

Lightfoot, D. W. (2015). How to trigger elements of I-languages. In A. Gallego & D. Ott (Eds.), *Fifty years later: Reflections on Chomsky's "Aspects"* (pp. 175–185). MIT Working Papers in Linguistics 77. Cambridge, MA: MIT Working Papers in Linguistics.

Lightfoot, D. W. (2016). [Review of Emonds & Faarlund 2014]. *Language, 92*(2), 474–477.

Lightfoot, D. W. (2017a). Acquisition and learnability. In A. Ledgeway & I. G. Roberts (Eds.), *Cambridge handbook of historical syntax* (pp. 381–400). Cambridge: Cambridge University Press.

Lightfoot, D. W. (2017b). Discovering new variable properties without parameters. In S. Karimi & M. Piattelli-Palmarini (Eds.), Parameters; special issue, *Linguistic Analysis*, *41*(3–4), 409–444.

Lightfoot, D. W. (2017c). Imperfect transmission and discontinuity. In A. Ledgeway & I. G. Roberts (Eds.), *Cambridge handbook of historical syntax* (pp. 515–533). Cambridge: Cambridge University Press.

Lightfoot, D. W. (2017d). Restructuring. In A. Ledgeway & I. G. Roberts (Eds.), *Cambridge handbook of historical syntax* (pp. 113–133). Cambridge: Cambridge University Press.

Lightfoot, D. W. (2018). Nothing in syntax makes sense except in the light of change. In A. Gallego & R. Martin (Eds.), *Language, syntax, and the natural sciences* (pp. 224–240). Cambridge: Cambridge University Press.

Lightfoot, D. W. & Havenhill, J. (Eds.). (2019). *Variable properties in language: Their nature and acquisition.* Washington, DC: Georgetown University Press.

Lightfoot, D. W. & Westergaard, M. (2007). Language acquisition and language change: Interrelationships. *Language and Linguistics Compass*, *1*(5), 396–416.

Lobeck, A. (1995). *Ellipsis: Functional heads, licensing, and identification.* Oxford: Oxford University Press.

Longobardi, G. (2001). Formal syntax, diachronic minimalism, and etymology: The history of French *chez. Linguistic Inquiry*, *32*(2), 275–302.

Longobardi, G. & Guardiano, C. (2017). Phylogenetic reconstruction in syntax: The parametric comparison method. In S. Karimi & M. Piattelli-Palmarini (Eds.), Parameters; special issue, *Linguistic Analysis*, *41*(3–4), 241–271.

Longobardi, G., Guardiano, C., Silvestri, G., Boattini, A., & Ceolin, A. (2013). Toward a syntactic phylogeny of modern Indo-European languages. *Journal of Historical Linguistics*, *3*(1), 122–152.

Mayberry, R. I. & R. Kluender (2018). Rethinking the critical period for language: New insights into an old question from American Sign Language. *Bilingualism: Language and Cognition*, *21*(5), 886–905.

Mayr, E. (1942). *Systematics and the origin of species.* New York: Columbia University Press.

McWhorter, J. H. (2009). What else happened to English? A brief for the Celtic hypothesis. *English Language and Linguistics*, *13*, 163–191.

McWhorter, J. H. (2016). Too good to be true: English is not Norse. *Language Dynamics and Change, 6*(1), 34–36.

Meisel, J. (2011). Bilingual language acquisition and theories of diachronic change: Bilingualism as cause and effect of grammatical change. *Bilingualism: Language and Cognition, 14*, 121–145.

Mitchener, W. G. & Nowak, M. A. (2004). Chaos and language. *Proceedings of the Royal Society of London B, 271*, 701–704.

Montrul, S. (2008). *Incomplete acquisition in bilingualism: Re-examining the age factor.* Amsterdam: Benjamins.

Musso, M., Moro, A., Glauche, V., Rijntjes, M., Reichenbach, J., Büchel, C., & Weiller, C. (2003). Broca's area and the language instinct. *Nature Neuroscience, 6*, 774–781.

Narayanan, H. & Niyogi, P. (2013). Language evolution, coalescent processes, and the consensus problem on a social network. Unpublished manuscript. http://faculty.washington.edu/harin/LangEvol.pdf.

Nelson, M. J., El Karoui, I., Giber, K., Yang, X., Cohen, L., Koopman, H., Cash, S. S., Naccache, L., Hale, J. T., Pallier, C., & Dehaene, S. (2017). Neurophysiological dynamics of phrase-structure building during sentence processing. *Proceedings of the National Academy of Sciences, 114*(18), E3669–E3678. https://doi.org/10.1073/pnas.1701590114.

Newmeyer, F. J. (2004). Against a parameter-setting approach to language variation. *Linguistic Variation Yearbook, 4*, 181–234.

Newmeyer, F. J. (2017). Where, if anywhere, are parameters? A critical historical overview of parametric theory. In C. Bowern, L. Horn, & R. Zanuttini (Eds.), *On looking into words (and beyond)* (pp. 547–569). Berlin: Language Science Press. https://doi.org/10.5281/zenodo.495465.

Newport, E. L. (1998). Reduced input in the acquisition of signed languages: Contributions to the study of creolization. In M. DeGraff (Ed.), *Language creation and change: Creolization, diachrony, and development* (pp. 161–178). Cambridge, MA: MIT Press.

Niyogi, P. (2006). *The computational nature of language learning and evolution.* Cambridge, MA: MIT Press.

Niyogi, P. & Berwick, R. (1995). *The logical problem of language change.* AI Memo 1516. Cambridge, MA: Artificial Intelligence Laboratory and Center for Biological Learning, Massachusetts Institute of Technology. http://hdl.handle.net/1721.1/7196.

Niyogi, P. & Berwick, R. (1997). A dynamical systems model of language change. *Complex Systems, 11,* 161–204.

Niyogi, P. & Berwick, R. (2009). The proper treatment of language acquisition and change in a population setting. *Proceedings of the National Academies of Science, 106*(25), 10124–10129.

Nunes, J. (2004). *Linearization of chains and sidewards movement.* Cambridge, MA: MIT Press.

Omaki, A. & Lidz, J. (2015). Linking parser development to acquisition of syntactic knowledge. *Language Acquisition, 22,* 158–192. https://doi.org/10.1080/10489223.2014.943903.

O'Neil, W. (1978). The evolution of the Germanic inflectional systems: A study in the causes of language change. *Orbis, 27*(2), 248–285.

Paul, H. (1877). Die Vocale der Flexions- und Ableitungssilben in den ältesten germanischen Dialecten. *Beiträge zur Geschichte der deutschen Sprache und Literatur, 4,* 314–475.

Paul, H. (1880). *Prinzipien der Sprachgeschichte.* Tübingen, Germany: Niemeyer.

Pesetsky, D. (1991). Zero syntax II: An essay on infinitives. Unpublished manuscript, Massachusetts Institute of Technology.

Pesetsky, D. (1995). *Zero syntax.* Cambridge, MA: MIT Press.

Peyraube, A. (1985). Les structures en *ba* en Chinois mediéval et moderne. *Cahiers de Linguistique—Asie Orientale, 14*(2), 193–213.

Phillips, C. (2003a). Linear order or constituency. *Linguistic Inquiry, 34*(1), 37–90.

Phillips, C. (2003b). Syntax. In L. Nadel (Ed.), *Encyclopedia of cognitive science* (Vol. 4, pp. 319–329). London: Nature Publishing Group.

Piattelli-Palmarini, M. (Ed.). (1980). *Language and learning: The debate between Jean Piaget and Noam Chomsky.* London: Routledge and Kegan Paul.

Piattelli-Palmarini, M. & Berwick, R. C. (Eds.). (2013). *Rich languages from poor inputs.* Oxford: Oxford University Press.

Pintzuk, S. (2002). Verb-object order in Old English: Variation as grammatical competition. In D. W. Lightfoot (Ed.), *Syntactic effects of morphological change* (pp. 276–299). Oxford: Oxford University Press.

Poletto, C. (2018). The twists and shakes of PRO Drop: On the licensing of null topics in Old Italian varieties. Paper presented at 20th Diachronic Generative Syntax meeting, DiGGS '20, York University, UK.

Potsdam, E. (1997). NegP and subjunctive complements in English. *Linguistic Inquiry, 28*(3), 533–541.

Preminger, O. (2014). *Agreement and its failures.* Linguistic Inquiry Monographs 68. Cambridge, MA: MIT Press.

Prigogine, I. & Stengers, I. (1997). *The end of certainty: Time, chaos, and the new laws of nature.* New York: The Free Press.

Richards, N. (2013). Lardil "case stacking" and the timing of case assignment syntax. *Syntax, 16*(1), 42–76.

Rizzi, L. (1978). Violations of the wh-island constraint in Italian and the Subjacency Condition. In C. Dubuisson, D. W. Lightfoot, & Y.-C. Morin (Eds.), *Montreal Working Papers in Linguistics,* 11, 155–190.

Rizzi, L. (1982). *Issues in Italian syntax.* Dordrecht, the Netherlands: Foris.

Rizzi, L. (1990). *Relativized minimality.* Cambridge, MA: MIT Press.

Rizzi, L. (2017). On the format and locus of parameters: The role of morphosyntactic features. In S. Karimi & M. Piattelli-Palmarini (Eds.), Parameters; special issue, *Linguistic Analysis, 41*(3–4), 159–192.

Rizzi, L. & Cinque, G. (2016). Functional categories and syntactic theory. *Annual Review of Linguistics, 2*(1), 139–163.

Roberts, I. G. (2007). *Diachronic syntax.* Oxford: Oxford University Press.

Roberts, I. G. (2017). Inertia. In A. Ledgeway & I. G. Roberts (Eds.), *Cambridge handbook of historical syntax* (pp. 425–445). Cambridge: Cambridge University Press.

Roberts, I. G. & Holmberg, A. (2010). Introduction: Parameters in minimalist theory. In T. Biberauer, A. Holmberg, I. G. Roberts, & M. Sheehan, *Parametric variation: Null subjects in minimalist theory* (pp. 1–57). Cambridge: Cambridge University Press.

Roberts, I. G. & Roussou, A. (2003). *Syntactic change: A minimalist approach to grammaticalisation.* Cambridge: Cambridge University Press.

Sakas, W. & Fodor, J. D. (2001). The structural triggers learner. In S. Bertolo (Ed.), *Language acquisition and learnability* (pp. 172–233). Cambridge: Cambridge University Press.

Sandler, W., Meir, I., Padden, C., & Aronoff, M. (2005). The emergence of grammar: Systematic structure in a new language. *Proceedings of the National Academy of Sciences, 102,* 2661–2665.

Sapir, E. (1925). Sound patterns in language. *Language, 1*(2), 37–51.

Senghas, A., Kita, S., & Özyürek, A. (2004). Children creating core properties of language: Evidence from an emerging sign language in Nicaragua. *Science, 305,* 1779–1782.

Singleton, J. L. & Newport, E. L. (2004). When learners surpass their models: The acquisition of American Sign Language from impoverished input. *Cognitive Psychology, 49,* 370–407.

Smith, N.V. & Tsimpli, I.-M. (1995). *The mind of a savant: Language learning and modularity.* Oxford: Blackwell.

Snyder, W. (2007). *Child language: The parametric approach.* Oxford: Oxford University Press.

Stokoe, W. (1960). Sign language structure: An outline of the visual communication systems of the American deaf. *Studies in linguistics: Occasional Papers* (No. 8). Buffalo: Department of Anthropology and Linguistics, University of Buffalo.

Sybesma, R. (1999). *The Mandarin VP.* Dordrecht, the Netherlands: Kluwer.

Tattersall, I. (2016, August 18). At the birth of language. *New York Review of Books,* 27–28.

Thom, R. (1972). *Stabilité structurelle et morphogénèse: Essai d'une théorie générale des modèles.* Reading, MA: W. A. Benjamin.

Thoms, G. (2019, January 4). On the uncommon emergence of preposition stranding. Presentation, ninety-third annual meeting of the Linguistic Society of America.

Thornton, R. & Wexler, K. (1999). *Principle B, VP ellipsis, and interpretation in child grammar.* Cambridge, MA: MIT Press.

Tinbergen, N. (1957). *The herring gull's world.* Oxford: Oxford University Press.

Trubetskoy, N. (1939). Gründzuge der Phonologie. In *Travaux du Cercle Linguistique de Prague* 7.

Trudgill, P. (2002). *Sociolinguistic variation and change.* Washington, DC: Georgetown University Press.

Trudgill, P. (2016). Norsified English or anglicized Norse? *Language Dynamics and Change, 6*(1), 46–48.

Valian, V. & Coulson, S. (1988). Anchor points in language learning: The role of marker frequency. *Journal of Memory and Language, 27*(1), 71–86.

van Craenenbroeck, J. & Merchant, J. (2013). Ellipsis phenomena. In M. den Dikken (Ed.), *Cambridge handbook of generative syntax* (pp. 701–745). Cambridge: Cambridge University Press.

van Gelderen, E. (2011). *The linguistic cycle: Language change and the language faculty.* Amsterdam: Benjamins.

van Kemenade, A. (1987). *Syntactic case and morphological case in the history of English.* Dordrecht, the Netherlands: Foris.

van Riemsdijk, H. C. (1978). *A case study in syntactic markedness.* Dordrecht, the Netherlands: Foris.

Vergnaud, J.-R. (2008). Letter to Noam Chomsky and Howard Lasnik on "Filters and control." In R. Freidin, C. P. Otero, & M.-L. Zubizarreta (Eds.), *Foundational issues in linguistic theory: Essays in honor of Jean-Roger Vergnaud* (pp. 3–16). Cambridge, MA: MIT Press. (Original letter sent 1977.)

Vikner, S. (1995). *Verb movement and expletive subjects in the Germanic languages.* Oxford: Oxford University Press.

Visser, F. T. (1963–1973). *An historical syntax of the English language.* Leiden, the Netherlands: Brill.

Walkden, G. (2012). Against inertia. *Lingua, 122,* 891–901.

Walkden, G. (2017, November 1). Preposition stranding in Early West Germanic. Presentation, LinG Colloquium, Georg-August-Universität Göttingen.

Wallenberg, J. C. (2013, August 2). A unified theory of stable variation, syntactic optionality, and syntactic change. Presentation, fifteenth Diachronic Generative Syntax conference (DiGS 15), University of Ottawa.

Wallenberg, J. C. (2016). Extraposition is disappearing. *Language, 92*(4), e237–e256.

Warner, A. (1995). Predicting the progressive passive: Parametric change within a lexicalist framework. *Language, 71,* 533–57.

Weiner, J. (1994). *The beak of the finch: A story of evolution in our time.* New York: Knopf.

Weiner, J. (1995). Evolution made visible. *Science, 267,* 30–33.

Weinreich, U., Labov, W., & Herzog, M. I. (1968). Empirical foundations for a theory of language change. In W. P. Lehmann & Y. Malkiel (Eds.), *Directions for historical linguistics: A symposium* (pp. 95–188). Austin: University of Texas Press.

Westergaard, M. (2009). Usage-based vs. rule-based learning: The acquisition of word order in *wh-* questions in English and Norwegian. *Journal of Child Language, 36*(5), 1023–1051.

Westergaard, M. (2014). Linguistic variation and micro-cues in first language acquisition. *Linguistic Variation, 14,* 26–45.

Westergaard, M. (2017). Gradience and gradualness vs abruptness. In A. Ledgeway & I. G. Roberts (Eds.), *Cambridge handbook of historical syntax* (pp. 446–466). Cambridge: Cambridge University Press.

Whitman, J. (2001). Relabelling. In S. Pintzuk, G. Tsoulas, & A. Warner (Eds.), *Diachronic syntax: Models and mechanisms* (pp. 220–238). Oxford: Oxford University Press.

Whitman, J. & Paul, W. (2005). Reanalysis and conservancy of structure in Chinese. In M. Batllori, M.-L. Hernanz, C. Picallo, & F. Roca (Eds.), *Grammaticalization and parametric variation* (pp. 82–94). Oxford: Oxford University Press.

Williams, E. (1978). Across-the-board rule application. *Linguistic Inquiry, 9*(1), 31–43.

Yang, C. (2002). *Knowledge and learning in natural language.* Oxford: Oxford University Press.

Yang, C. (2006). *The infinite gift: How children learn and unlearn the languages of the world.* New York: Scribner's.

Yang, C. (2016). *The price of linguistic productivity: How children learn to break the rules of language.* Cambridge, MA: MIT Press.

Yang, C. & Roeper, T. (2011). Minimalism and language acquisition. In C. Boeckx (Ed.), *Oxford handbook of linguistic Minimalism.* Oxford: Oxford University Press.

Zagona, K. (1988). Proper government of antecedentless VP in English and Spanish. *Natural Language and Linguistic Theory, 6*, 95–128.

Zimmerman, M. (2012). *The evolution of expletive subject pronouns in French.* Unpublished PhD dissertation, University of Konstanz.

Zwicky, A. M. (1994). What is a clitic? In J. Nevis, B. Joseph, D. Wanner, & A. M. Zwicky (Eds.), *Clitics: A comprehensive bibliography* (pp. xii–xx). Amsterdam: Benjamins.

Zwicky, A. M. & Pullum, G. K. (1983). Cliticization vs. inflection: English *n't. Language, 59*(3), 502–513.

Index